INTERNATIONAL ENERGY AGENCY

Energy and Climate Change

an **IEA**
Source-Book
for **Kyoto**
and beyond

INTERNATIONAL ENERGY AGENCY
9, RUE DE LA FÉDÉRATION, 75739 PARIS CEDEX 15, FRANCE

The International Energy Agency (IEA) is an autonomous body which was established in November 1974 within the framework of the Organisation for Economic Co-operation and Development (OECD) to implement an international energy programme.

It carries out a comprehensive programme of energy co-operation among twenty-four* of the OECD's twenty-nine Member countries. The basic aims of the IEA are:
- To maintain and improve systems for coping with oil supply disruptions;
- To promote rational energy policies in a global context through co-operative relations with non-Member countries, industry and international organisations;
- To operate a permanent information system on the international oil market;
- To improve the world's energy supply and demand structure by developing alternative energy sources and increasing the efficiency of energy use;
- To assist in the integration of environmental and energy policies.

* *IEA Member countries: Australia, Austria, Belgium, Canada, Denmark, Finland, France, Germany, Greece, Hungary, Ireland, Italy, Japan, Luxembourg, the Netherlands, New Zealand, Norway, Portugal, Spain, Sweden, Switzerland, Turkey, the United Kingdom, the United States. The European Commission also takes part in the work of the IEA.*

ORGANISATION FOR ECONOMIC CO-OPERATION AND DEVELOPMENT

Pursuant to Article 1 of the Convention signed in Paris on 14th December 1960, and which came into force on 30th September 1961, the Organisation for Economic Co-operation and Development (OECD) shall promote policies designed:

- to achieve the highest sustainable economic growth and employment and a rising standard of living in Member countries, while maintaining financial stability, and thus to contribute to the development of the world economy;
- to contribute to sound economic expansion in Member as well as non-member countries in the process of economic development; and
- to contribute to the expansion of world trade on a multilateral, non-discriminatory basis in accordance with international obligations.

The original Member countries of the OECD are Austria, Belgium, Canada, Denmark, France, Germany, Greece, Iceland, Ireland, Italy, Luxembourg, the Netherlands, Norway, Portugal, Spain, Sweden, Switzerland, Turkey, the United Kingdom and the United States. The following countries became Members subsequently through accession at the dates indicated hereafter: Japan (28th April 1964), Finland (28th January 1969), Australia (7th June 1971), New Zealand (29th May 1973), Mexico (18th May 1994), the Czech Republic (21st December 1995), Hungary (7th May 1996), Poland (22nd November 1996) and the Republic of Korea (12th December 1996). The Commission of the European Communities takes part in the work of the OECD (Article 13 of the OECD Convention).

© OECD/IEA, 1997
Applications for permission to reproduce or translate all or part of this publication should be made to:
Head of Publications Service, OECD
2, rue André-Pascal, 75775 PARIS CEDEX 16, France.

FOREWORD

Energy production and use are a major part of the climate change problem. Therefore, energy and the way we use it must be a major part of the solution. The purpose of this book is to provide a deeper understanding of energy trends and related carbon dioxide emissions. It contains results from recent IEA analyses which will be useful for the climate change negotiations in Kyoto in December, 1997 (COP-3) and for the work that follows.

Beginning with the IEA Statement on the Energy Dimension of Climate Change, the book focuses on prospects for carbon dioxide emissions and on ways to limit and reduce them. The most recent IEA projections of world energy demand from our 1996 World Energy Outlook are presented by reference to trends in energy-related services, namely electricity, mobility and heat. The study discusses different policies to reduce the growth of emissions, their advantages and limitations and the political difficulties that may be involved. It shows how technological developments can lead to emissions reductions in the future as more advanced power plants, industrial processes, buildings and vehicles replace existing stocks. The book also discusses ways in which developed and developing countries can work together to resolve climate change issues.

Robert Priddle
Executive Director

TABLE OF CONTENTS

FOREWORD	3
EXECUTIVE SUMMARY	7

1 IEA STATEMENT ON THE ENERGY DIMENSION OF CLIMATE CHANGE.... 13

2 CO_2 EMISSIONS IMPLICATIONS OF THE IEA 1996 *"WORLD ENERGY OUTLOOK"* CAPACITY CONSTRAINTS CASE........................... 41
  Introduction 41
  The Outlook for Energy Demand: An Energy-Related Services Analysis........... 43
  The Outlook for CO_2 Emissions 48
  Summary 52

3 CO_2 EMISSIONS IMPLICATIONS OF THE IEA 1996 *"WORLD ENERGY OUTLOOK"* ENERGY SAVINGS CASE 53
  Energy-Related CO_2 Emissions in the Energy Savings Case.................... 53
  Sectoral Variations 57
  Regional Variations 59
  How Energy Savings Might be Achieved in the Energy Savings Case 61
  Summary 61

4 REALISING ENERGY SAVINGS POTENTIAL TO 2010: THE POLICY RESPONSE .. 63
  Introduction 63
  Policies and Measures by Sector 73
  Summary 82

5 LONGER-TERM FACTORS AFFECTING CLIMATE CHANGE POLICIES.......... 85
  Introduction 85
  Technological Development.. 85
  Factors Affecting Energy Demand 86
  Possible Developments in Energy Transformation 88
  The Future for Fossil Energy Supplies................................... 91
  Climate Change Policies and Economic Growth 93
  Summary 95

EXECUTIVE SUMMARY

Climate change is a major global issue with profound implications for the way the world produces and consumes energy. IEA Ministers have repeatedly renewed their commitment to adopt policies aimed at more environmentally friendly forms of energy use and production. Climate response strategies being developed by IEA countries need to be based on a solid understanding of energy systems, an awareness of the magnitude of the policy challenge and an assessment of practical options for cost-effective action by governments. This report analyses historical trends and IEA projections of energy-related CO_2 emissions and assesses policy options for reducing these emissions.

The official IEA Statement on the Energy Dimension of Climate Change, adopted by the IEA Governing Board in January 1997, constitutes the first chapter. The statement represents the agreed views of IEA energy ministers on key energy-related aspects of the Berlin Mandate, which was adopted in 1995, under the United Nations Framework Convention on Climate Change (UNFCCC). The objective of this statement is to summarise for all participants in the UNFCCC process the energy dimension of the climate change issue. It is republished here on the conviction that it would be directly useful in the run-up to the third meeting of the Conference of the Parties (COP-3) under the UNFCCC and in any subsequent negotiation. The rest of the report focuses on prospects for emissions and the policy challenge. It draws largely on the most recent IEA projections of world energy demand and related CO_2 emissions set out in the 1996 *World Energy Outlook* (WEO). The last chapter treats longer-term issues relevant to climate change policy in the period beyond 2010.

CO_2 is the single most important anthropogenic greenhouse gas; fossil fuel production and use cause about three quarters of man-made CO_2 emissions. Other energy-related greenhouse gases include nitrogen oxides (primarily from the burning of wood as fuel), methane (from the production, transportation and use of natural gas and coal) and other precursors to tropospheric ozone. The energy sector (from primary energy extraction to end-uses) is the major source of CO_2 emissions.

Energy demand and related CO_2 emissions have increased substantially over the past 50 years, along with economic development and demographic growth. Most of that growth in energy demand and CO_2 emissions has occurred in OECD countries, but it has also occurred increasingly in regions outside OECD over the past two decades. Clear differences exist between use-patterns in transportation (demand for mobility), electricity generation, electricity consumption, and other stationary end-uses of fossil fuels. Historically, demand for electricity and transportation fuel closely followed economic growth. Fossil fuel demand for stationary heat purposes has been more sensitive to energy price movements, particularly during the 1973 and 1979 oil shocks, and has levelled out in recent years.

The different dynamics of the main energy services (mobility, electricity use and heat) result from different infrastructures, technologies, lifetimes of capital stocks, and the nature and behaviour of decision makers. Attempts to restrain consumption of energy services meet resistance, but it is often possible to obtain reductions in consumption indirectly, through the introduction of more climate-friendly technologies. Many items of capital stock have long lifetimes. It is important to take full advantage of the turnover of these capital items in order to introduce these new technologies.

Past trends are expected to persist in the medium term, in the absence of a sharp downturn in economic activity or major policy initiatives to curb energy demand and emissions. In the Capacity Constraints or "Business-As-Usual" Case of the World Energy Outlook (which assumes no major policy intervention), world demand for primary energy continues to grow steadily, rising by 46% between 1993 and 2010 — the same growth rate as in the period from 1971 to 1993. This increase in demand is met almost entirely by fossil fuels, which are projected to account for almost 90% of total primary energy use in 2010. Global CO_2 emissions are also projected to grow substantially — by 50% between 1990 and 2010. The bulk of increases in both demand and emissions is expected to occur in non-OECD countries.

The Energy Savings Case of the *World Energy Outlook* assumes cost-effective, "no-regrets"[1] reductions in energy use and related CO_2 emissions, amounting to at least an 8% saving in world energy use and a 9% reduction in emissions. Environmental costs and benefits are not included in the calculations. It is further assumed that energy prices will remain broadly flat until 2010, in contrast with the Capacity Constraints Case, which foresees rising prices from 2000. The power sector is expected to offer the single largest source of emissions reductions, accounting for around one half of the world total, followed by stationary uses. The OECD accounts for more than 50% of the total emissions reduction. This otherwise untapped potential is explained by the existence of barriers to energy-efficient choices, for example: lack of information, strong preferences for alternative expenditures, regulatory impediments and limitations on income and credit.

It is important to distinguish among three main types of energy efficiency policies:

■ those that provide information on how to save energy;

■ those that limit consumer choice or require changes in consumer behaviour;

■ those that involve changes in energy prices.

---

1 Cost-effective, no-regret, here means showing a net cash benefit to the end-user without taking account of any environmental benefit.

Several other policies can also bring about energy saving, for example: industrial restructuring, especially in the Former Soviet Union and Eastern and Central Europe (FSU/ECE), or transport policies, such as congestion charges that reduce traffic rather than divert it.

The potential for realisable no-regret, cost-effective energy saving is very uncertain. Current estimates by the Intergovernmental Panel on Climate Change (IPCC) suggest a range of 10%-30% gains on baseline trends over the next two to three decades. This potential will be realised only if policies and measures can be found and implemented to overcome existing barriers. Political difficulties will limit the size of no-regret savings. Policies to achieve significant energy savings are likely in practice to involve both low to medium economic cost and low to medium political pain.

IEA Member countries have implemented a wide range of energy efficiency policies over the last twenty years. Some have introduced measures to raise the fuel efficiencies of new cars, (the United States, Canada, Japan and Germany). Many countries have introduced tighter building regulations and stricter energy efficiency standards for electrical appliances during the 1990s. Other policies have concentrated on providing better information. The energy savings achieved in IEA regions so far are much less than those that would be required to reduce CO_2 emissions in 2010 below their 1990 levels.

Governments in some countries may find even low-cost policies difficult to implement politically, e.g. removal of subsidies for energy production and pricing, or exposure of domestic producers to competition from imports. Electorates will need to be persuaded on the basis of broader social benefits, such as reduced urban pollution and traffic congestion, and satisfied on issues of equity.

Low-cost policies include removal of the market distortions of energy subsidies and probably also the institution of regulatory reforms (particularly in the electricity and natural gas sectors). Other policies and measures, such as voluntary agreements, relatively undemanding energy efficiency performance standards and energy efficiency information dissemination, technical assistance and education/training programmes, are also generally low-cost. More ambitious targets for reducing emissions will involve higher costs and may require carbon taxes, tradeable permits and more stringent regulations on fuel use and emissions.

Emissions reductions are likely to be more difficult to achieve in the transport and electricity end-use sectors than in stationary end-uses or electricity generation. In both the former sectors, the fossil fuel cost element is only a small part of the final fuel price to the consumer; fuel costs are only a part of the unit cost of providing the service of mobility or of electrical appliances. In addition, the price elasticities of demand for these fuels are low. Very large taxes on the basic fuel cost would be required to have any significant effect on fuel use. Possible alternative approaches include:

- In the transport sector, policy packages, including congestion charges in cities, motorway tolls, vehicle/road taxes differentiated by size of vehicle, creation of expectations of increases in the cost of travel by road and the provision of alternative public transport modes. These are policies for transport rather than for energy use, but they would also save energy and reduce CO_2 emissions.

- In the electricity sector, policies that encourage the design and production of more energy-efficient appliances, coupled with minimum energy efficiency standards. These policies could help offset the low energy price responsiveness of this sector.

* Business-as-usual energy projections indicate that CO_2 concentrations will continue to rise and eventually exceed twice the pre-Industrial Revolution level, unless substantial new and effective CO_2 emission abatement policies are implemented.

* Economic instruments are potentially the least costly way to achieve significant reductions in CO_2 emissions, particularly if co-ordinated internationally. An internationally agreed economic instrument, such as an emissions trading system, could complement measures implemented by individual countries at the national level.

* Climate change policies agreed in Kyoto in 1997 could affect some capital stocks of equipment, buildings and vehicles by 2010.

* Technology development has the greatest potential in the longer term for reducing the unit costs of CO_2 emission abatement.

* A package of climate change policies is likely to reduce coal consumption and boost natural gas demand, compared with a "business-as-usual" projection; oil demand may be little affected.

* Nuclear power will be favoured in some countries, but will be subject to public acceptability, safety, cost and environmental factors.

* Renewable energy sources are also likely to gain an advantage over fossil fuels under climate change policies, but they will still be costly to produce, unsightly and environmentally problematical.

* Fossil fuel reserve estimates are extremely uncertain, although these reserves are extensive (in their various forms). However, the range of current estimates suggests that production of conventional oil could peak in the period 2000-2020, assuming that current consumption trends continue. Natural gas supply is expected to be sufficient to meet growing demand beyond 2020, but may peak before 2035.

- All countries will need to co-operate to address the climate change threat. Annex I countries[2] have agreed to take the lead in assuming quantified emissions limitation or reduction commitments. Developing countries already have reporting, inventory and mitigation commitments. Joint Implementation, with credits accruing to participating countries on a mutually agreed basis, can provide a market-based monetary incentive to reduce emissions cost effectively and to transfer efficient, climate-friendly technologies to developing countries. Developed countries would bear a substantial part of the incremental cost of the reductions.

---

2   Annex I countries comprise the countries of the OECD and FSU/ECE.

1

INTERNATIONAL ENERGY AGENCY STATEMENT ON THE ENERGY DIMENSION OF CLIMATE CHANGE

(This statement was approved by all IEA Member countries* at the Governing Board Meeting in February 1997 and endorsed by IEA Energy Ministers at their meeting in May 1997.)

INTRODUCTION

Climate change is a major global issue ...

Climate change is a major global issue with profound implications for the way the world produces and consumes energy. The Second Assessment Report of the Intergovernmental Panel on Climate Change (IPCC) states that in spite of remaining uncertainty "the balance of evidence suggests that there is a discernible human influence on global climate".

... and IEA Ministers are committed to addressing it

On several occasions, IEA Ministers have expressed commitment to adopt policies aimed at more environmentally friendly forms of energy consumption and production, and sustainable economic development.

This document offers a perspective from energy ministries of IEA Member countries on key energy-related aspects of the Berlin Mandate under the United Nations Framework Convention on Climate Change (UNFCCC).

---

* The International Energy Agency participating countries are: Australia, Austria, Belgium, Canada, Denmark, Finland, France, Germany, Greece, Hungary, Ireland, Italy, Japan, Luxembourg, the Netherlands, New Zealand, Norway (by special agreement), Portugal, Spain, Sweden, Switzerland, Turkey, the United Kingdom and the United States. The European Commission takes part in the work of the IEA.

This statement summarises the main driving forces related to greenhouse gas emissions from energy

The objective of this statement is to summarise for all participants in the UNFCCC process the energy dimension of the climate change issue in a manner that is directly useful in the run-up to the third meeting of the Conference of the Parties (COP-3) under the UNFCCC.

It describes the global context and continued trends in CO_2 emissions from the dynamic perspective of energy-related services. It also provides insights on key parameters to reduce energy-related greenhouse gas (GHG) emissions in a cost-effective, practical manner.

GLOBAL CONTEXT AND TRENDS IN ENERGY-RELATED CO_2 EMISSIONS

While the UNFCCC process covers all countries ...

At the United Nations Conference on Environment and Development in Rio, a balance was struck between the equally pressing global issues of environment and development. It was recognised there that countries had a right to, and should promote, sustainable development, alongside the need to tackle global environmental issues.

... only Annex I Parties are committed to aim to return their greenhouse gas emissions to 1990 levels by the year 2000

The United Nations Framework Convention on Climate Change adopted as a principle that Parties should protect the climate system ..."*on the basis of equity and in accordance with their common but differentiated responsibilities and respective capabilities. Accordingly, the developed country Parties should take the lead in combatting climate change and the adverse effects thereof* (Article 3)". While the aim, contained in Article 4.2, to return emissions to 1990 levels applies only to Annex I Parties, the UNFCCC also includes Article 4.1, which calls on all Parties, including developing country Parties, to formulate and implement programmes to mitigate climate change and facilitate adaptation to climate change.

The ultimate objective of the UNFCCC, as set out in Article 2, is to stabilise atmospheric GHG concentrations at non-dangerous levels. Analysis of global emission trends shows that the stabilisation of GHG concentrations in the atmosphere cannot be achieved by Annex I countries alone. In the long run, wider participation and efforts to limit emissions and enhance sinks will be required.

Globally, CO_2 is the main anthropogenic greenhouse gas

The UNFCCC calls for a comprehensive approach addressing all greenhouse gases, all sources and sinks, and both mitigation of and adaptation to climate change. The focus of this paper is on mitigation of CO_2, but extends to other greenhouse gases from energy. In the overall context of the Convention, other gases and emitting activities also need to continue to be addressed.

CO_2 is the single most important anthropogenic greenhouse gas; fossil fuel production and use represent about three quarters of man-made CO_2 emissions. Other energy-related greenhouse gases include CH_4 (from production, transportation and use of natural gas and coal), N_2O (primarily from fuel wood use) and other precursors to tropospheric ozone (O_3).

The energy sector (from primary energy extraction to end-uses) has been the major source of CO_2 build-up and can make a significant contribution to address the climate change problem.

Energy has been decisive in both economic development and increased CO_2 emissions from the burning of fossil fuels

Energy is an essential factor of economic activity. It contributes directly to meeting both basic and more sophisticated human needs, whether it takes the form of a primary energy source such as biomass fuel, coal, natural gas, oil, renewable energy sources, or a secondary or transformed form of energy such as refined oil products or electricity (based on fossil fuels, renewables or nuclear energy). Energy is both a traded commodity and an essential factor in moving goods that are traded internationally; as such it also contributes indirectly to economic growth.

Since the beginning of the industrial era, fossil energy has fuelled economic growth, leading to a sharp increase in greenhouse gas emission levels and their build-up in the atmosphere. Fossil fuels currently amount to 84% and 92% of *commercial* energy use in IEA countries and in the rest of the world, respectively.

The following graph from the IPCC shows that energy-related CO_2 emissions have increased substantially over the past 50 years, along with economic development and demographic growth, mostly in the OECD, but increasingly in regions outside OECD over the past two decades.

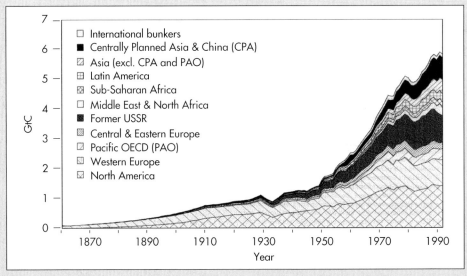

Figure 1.1
Global Energy-related CO_2 Emissions by Major World Region in billion tons of Carbon per Year

Source: IPCC 1996

Energy intensity has decreased over time

The close link between energy use and economic activity is by no means "one for one". On the contrary, the experience is that the amount of energy — non-commercial (e.g., fuelwood) and commercial (marketed fossil fuels and electricity) — required to produce output in terms of GDP (i.e., "energy intensity") tends to decrease over time.

In particular, shifts from non-commercial to commercial energy sources in the process of economic development and more efficient production processes have generally led to decreasing energy intensity. In this overall process, energy security and environmental protection have also promoted more rational energy uses. In brief, the general dynamics of energy intensity are influenced by three overlapping phases:

☐ a shift from non-commercial to commercial energy;

☐ an increase in efficiency of commercial energy use;

☐ a substitution of electricity for direct fossil fuel uses for many energy services.

Yet there remain differences among countries: some countries with a comparative advantage in energy resources have attracted energy-intensive activities, thus raising their energy intensity.

Figure 1.2
Evolution of Energy Intensity: an Illustration

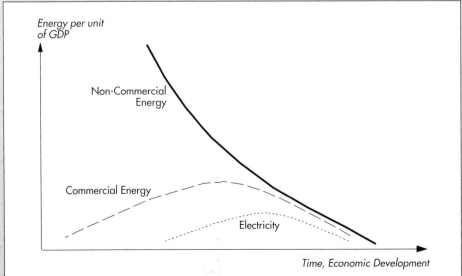

Energy intensity varies widely across regions and countries...

Physical factors such as geography and climate, population density, resource endowments, economic structure and the degree of competition in domestic energy markets influence current energy use across countries, through their effects on local economic activities, mobility needs, heating and cooling of buildings, and the availability and relative cost of energy sources. These local and national circumstances vary widely among IEA countries and are reflected in their level of CO_2 emissions per capita and per unit of GDP.

Furthermore, energy systems of IEA countries show variations in: the composition of primary energy supplies; the structure of electricity grids and other communication networks; the age of the system's components; the density of users; the degree of deregulation; and the division of ownership between the public and private sectors.

... and between energy services

Clear differences between patterns in transportation (demand for mobility), electricity generation, electricity consumption, and other stationary end-uses of fossil fuels point to the need to understand the specific features of each energy service. Without a disaggregated approach which recognises these differences, analysis may have limited relevance to practical response options.

The different dynamics of the main energy services (mobility, electricity use and heat) come from different infrastructures, lifetime of capital stocks, technologies, and the nature and behaviour of involved decision-makers (households versus private companies versus governments, whether at local or national level).

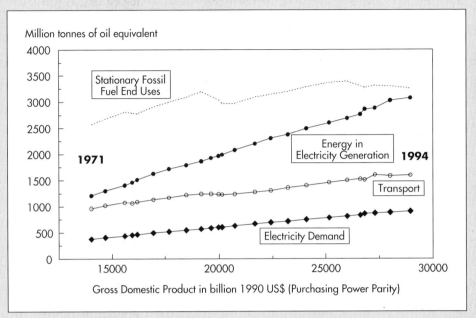

Figure 1.3
World Energy Demand: a Sectoral View, 1971-1994

Historically, electricity and transportation fuel demand have closely followed economic growth

Electricity use and energy use to fulfil mobility needs have closely followed world economic output. The above graph also shows the corresponding energy needed to produce end-use electricity (fossil fuels, nuclear, hydro and other renewable sources). The difference between energy use for electricity generation and demand represents losses from the transformation of primary energy into electricity, and the transportation and distribution of electricity. These losses should decrease somewhat over time as more efficient fossil fuel transformation technologies (gas turbines, combined heat and power generation) are installed.

Demand for heat from fossil fuels presents a more complex picture

Three major events have influenced fossil fuel demand for stationary heat purposes: the 1973 and 1979 oil shocks and the economic restructuring of centrally-planned European economies after 1989. Successful energy efficiency programmes, a shift towards service-oriented activities which require less energy to produce, as well as the relocation of some industrial activities to developing countries explain the stabilisation of heat-related fossil fuel demand in IEA countries as a whole.

Since the late seventies, most of the increase in stationary use of fossil fuels for heat services has taken place outside IEA countries. Economic development is driving the demand for fossil fuel-based heat, especially for industrial activities and also leads to the substitution of commercial fuels for non-marketed traditional fuels.

The following graph illustrates the evolution of energy demand related to mobility, electricity use and other stationary use of fossil fuels for IEA countries over the past 35 years. It shows in particular the stabilisation of fossil fuel demand for stationary end-uses other than electricity.

Figure 1.4
Energy Demand in IEA Countries: a Sectoral View, 1960-1994

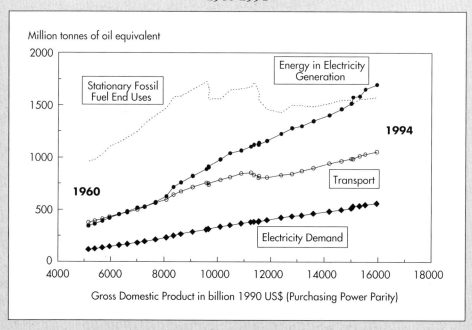

Future economic growth will likely trigger increasing energy-related CO_2 emissions in the absence of new measures ...

In the absence of effective action to address climate change, the IEA World Energy Outlook indicates that, within the range of cases examined, continuing economic growth is likely to lead to growth in energy-related CO_2 emissions in IEA countries over the next 15 years.

The overall carbon content of IEA countries' energy requirements has been decreasing over the past 20 years (see following graph), due in particular to the development of nuclear power in response to energy security concerns after the two oil shocks and to increasing use of natural gas, whose combustion contributes to less GHG emissions than oil and coal.

In the absence of specific responses, the observed rate of "de-carbonisation" of energy in IEA economies is unlikely to be maintained, since nuclear programmes have now been slowed or halted in most countries.

Renewable energy technologies other than hydro are not in most cases cost competitive with major

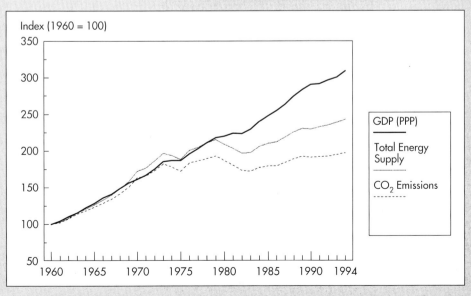

Figure 1.5
**Trends in GDP, Energy and CO_2
IEA Countries**

conventional energy sources. They have however undergone dramatic cost reductions and, should this trend continue, could further reduce the carbon content of energy. Their contribution is small today, yet current expectations on relative costs suggest the share of renewables will grow.

... especially in developing countries

Ongoing economic development in non-Annex I countries will also contribute to rising energy use and related CO_2 emissions, at a faster rate than in developed countries. However, while there is a wide variety of situations in non-Annex I countries with respect to the levels of economic development, energy use, and anthropogenic GHG emissions, on aggregate, per capita energy use and related GHG emissions will remain much lower than those of developed countries for some time to come.

DYNAMICS OF ENERGY SUPPLY AND DEMAND IN IEA COUNTRIES

Energy provides services ...

Climate change energy responses should be based on a clear understanding of the factors influencing energy demand and related CO_2 emissions. Energy is consumed not for itself but for the services it provides: mobility, heat, and electricity for a variety of end-uses.

These energy services are produced through a chain of technologies and infrastructures which influence the call for these services in the future and the way they are delivered. For example, cars require roads, but the availability of roads influences patterns of settlement, and thus gives rise to more demand for transportation.

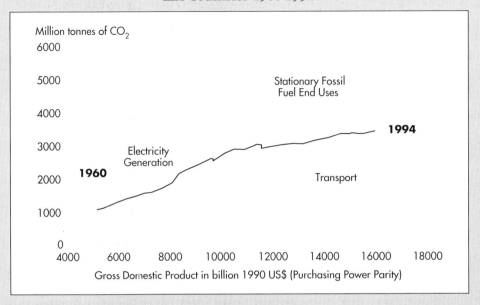

Figure 1.6
**Energy-Related CO_2 Emissions,
IEA Countries 1960-1994**

... that consumers may be reluctant to do without

The trends of CO_2 emissions in transportation, electricity and heat show some regularity, except for the reactions to major disruptions such as the two oil shocks (see graph above).

Attempts to constrain energy-related services typically encounter strong resistance from consumers, although reductions can be obtained indirectly through the introduction of end-use technologies which bring the same or better services in a more climate-friendly manner; changes in consumers' behaviour may be more feasible over the long term, as consumers become more attuned to the greenhouse effect.

Infrastructure limits the near-term flexibility of energy systems

The infrastructure which frames energy use (residential and commercial buildings, industrial "shells", roads, energy grids) is capital-intensive. Much of the currently installed energy-using capital stock, with the exception of the 1974-86 period associated with the two oil shocks, was designed and chosen in a context of relatively low energy prices, with little incentive to focus on energy efficiency improvements. Retiring capital stocks before their normal life will in many cases be costly; however, cost-effective reductions in energy use and emissions can be achieved by retrofits especially if carried out for other purposes. The rigidities inherent in existing infrastructure and associated inertia in individual behaviours require ongoing efforts to ensure that choices of infrastructure are taken with full knowledge of their implications on future GHG emissions.

Technological improvements are also constrained in the near-term

The enhanced use of best available technologies could help reduce energy requirements and CO_2 emissions within the constraint of current infrastructure. The barriers to the adoption of currently available and cost-effective technologies will need to be overcome. These include their perceived lower profitability (whenever their environmental benefits are not recognized), awareness of their potential and their reliability, the local knowledge and experience with their operation, and problems associated with the transfer of proprietary technologies. At the end, their ultimate contribution remains limited by their technical potential.

Along with the need to orient infrastructure towards more climate-friendly choices, the development of new technologies and processes for the future is an essential part of a longer term strategy to reduce greenhouse gas emissions. Longer term options exist for the various energy services, but action needs to be pursued vigorously so that more climate-friendly technologies can be assessed, developed and introduced into the market within the next few decades.

Price effects occur within these constraints ...

Energy prices affect the behaviour of energy users in the short run and influence long term infrastructure development and capital investment.

Because the three major services (mobility, electricity use and heat) are different in essence, as well as in terms of underlying infrastructures, sensitivity to the price of energy, and the capacity to react to price changes, vary widely.

... with major sectoral variations

Fossil fuel demand for heat as a whole shows more sensitivity to prices in the short run than the two other energy services (mobility and electricity). Energy-intensive industries, where energy expenditures represent a significant share of production cost, tend to manage their energy use more closely than less energy-intensive industries and non-industrial users.

Even though there have not been either upward or downward long-term trends in end-use prices, comparisons across IEA countries show that consistently higher energy prices have led to lower energy demand for related services.

Understanding the dynamics of complex energy systems requires considering the economic level of the country, the motivation of diverse actors and sectors, and other factors. Such understanding is the key to a cost-effective approach to CO_2 emissions reduction.

Mobility patterns reflect infrastructure ...

The mobility of goods and persons is an important component of economic activity in IEA countries. The absolute increase in energy use for transportation is due primarily to personal cars and road freight, and to a lesser extent to sea and air travel. In the overall demand for transportation, patterns of human settlements, population density, and income remain essential factors in explaining differences across IEA countries.

Large investments have been made in the past to support road-transportation, which in turn have shaped population settlements and mobility needs. Thus, in the short to medium term, mobility patterns are largely determined by historic decisions on pricing of transport fuels, cars, and on infra-structural choices (roads versus railroads, for instance). Given the inertia of existing infrastructure, mobility patterns can only be changed in the long run. In the near-term, efficiency improvements in cars and modal shifts (from private to public transportation) can help reduce energy demand for similar mobility needs.

All IEA countries have experienced a rather linear growth in energy demand for mobility, a combination of low specific consumption, higher number of cars, and increased usage. In North America, the introduction of an average fuel economy standard for car manufacturers has had a marked impact on the trend. The following graph indicates the average trend in energy demand for transportation for IEA countries over the 1960-1994 period, along with GDP growth, as well as the range of diverse situations, illustrated by two countries.

Figure 1.7
Energy in Road Transport: Range of Past Trends in IEA Countries, 1960-1994

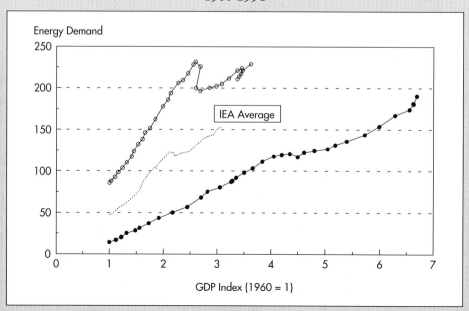

... while energy demand also reflects vehicle characteristics and use...

The energy required for any given pattern of mobility is affected by the average efficiency of vehicles which have much shorter lifetimes than transportation infrastructures: most of the effects of any policy or market change will be seen in on-road technology within at most 15 to 20 years (i.e., from technology design to market absorption). This suggests the need to orient research towards more advanced transportation technologies, from vehicles to infrastructure, in order to be able to cope with longer run constraints on CO_2 emissions.

... and end-user prices

Pricing of transportation has been largely related to the historical availability of domestic resources and the reliance on oil imports which differ across IEA countries.

There is clear evidence throughout IEA countries that fuel prices including taxes, taxes on cars, or charges for their use, as well as direct government

Figure 1.8
**Energy Demand in Transport per Unit of GDP and Prices
IEA Countries, 1994**

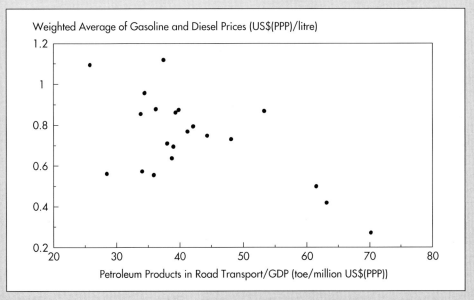

regulations on car efficiency, can have a strong influence on the intensity of energy demand in the transportation sector per unit of GDP, through the combination of vehicle choice (gasoline versus diesel versus alternative fuels), design (including car size and accessories), their use (vehicle management and annual mileage driven), and the modal split between public and private transportation.

Electricity demand reflects ongoing technological change in a wide variety of activities

Electricity provides the most convenient fuel for a wide range of end-uses, and is currently irreplaceable in many of them. The growth of electricity demand reflects this convenience, the general increase in energy demand, the development of new electricity uses as a result of technological innovation, and substitution away from fossil fuel end-uses. While there is constant improvement of energy efficiency in electric appliances, increasing levels of activities and needs have resulted in the observed increase in electricity demand.

Figure 1.9
Electricity Demand: Range of Past Trends in IEA Countries, 1960-1994

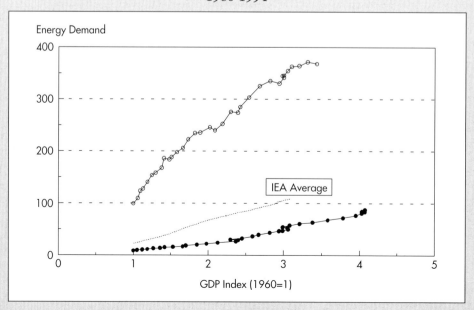

On the electricity production side, baseload generation concentrates on the most readily available supply sources under national circumstances, while peak-load typically uses oil products as the fuel at the margin.

Electricity demand appears to grow constantly with GDP...

The above graph indicates the average trend in electricity demand against GDP for IEA as well as an indication of the range across Member countries.

... with variations depending on the price environment and resource base

The electricity intensity of IEA economies and the rate of adoption of new electricity uses vary from one country to another with prices faced by industry, services and households. These prices depend on a number of factors such as the availability of domestic primary energy sources, the efficiency of power generation, market structure, etc.

Figure 1.10
**Electricity Demand per Unit of GDP and Prices
IEA Countries, 1994**

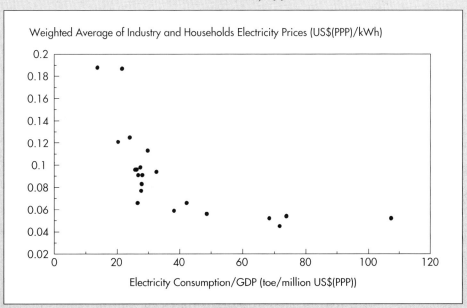

Stationary fossil fuel end-uses are more price-sensitive in the near term

Contrary to transportation and electricity demand, the demand for heat (i.e., stationary fossil fuel end-uses other than for electricity production) has shown significant drops following the oil shocks (see next graph, with an indication of the range among IEA countries). Another feature is the shift of heat demand trends thereafter, leading to a stabilisation of related CO_2 emissions from these end-uses, despite substantially lower oil prices since 1986 (see earlier graph: *Energy-related CO_2 emissions, IEA countries*).

Beyond the overall structural change towards energy services in IEA economies, the current situation was reached through a combination of high concern about security of supply (initially), specific policies, efforts of industries towards energy conservation and industrial relocation.

Since fossil fuels represent a significant share of costs of heat production, fossil fuel prices are a key parameter in production patterns. The 1994 cross-section observation of IEA member countries reveals a clear correlation between end-use prices and heat demand per unit of GDP (see following graph).

Figure 1.11
Stationary Fossil Fuel End Uses: Range of Past Trends in IEA Countries, 1960-1994

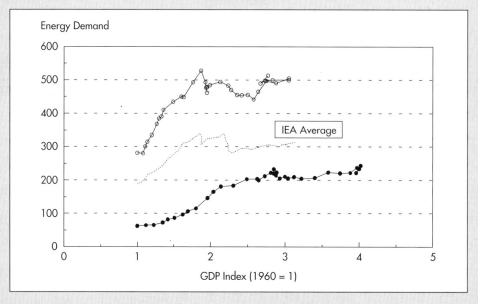

Nevertheless, many other factors affect the fossil fuel demand for heat services. For instance, despite low energy prices, some countries continue to achieve low energy intensity by improving efficiency through regulations and voluntary measures. This demonstrates that the trend shift in heat demand is influenced by policy actions.

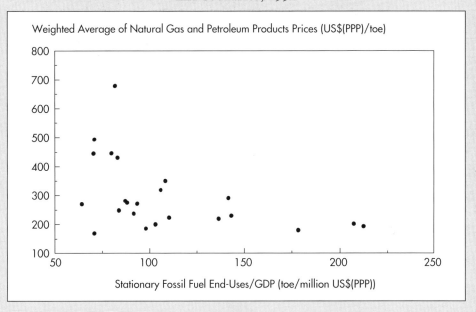

Figure 1.12
**Stationary Fossil Fuel Demand per Unit of GDP and Prices
IEA Countries, 1994**

Actions need to take full benefit of capital stock turnover

The extent and age of existing energy infrastructures constrain the rate at which cost-effective reductions in energy-related emissions can be achieved in the near/medium term. On the other hand, energy systems are constantly changing, as capital stocks are replaced and/or expanded along the full energy chain, from extraction to service provision.

In many cases, there is a large potential for energy efficiency improvement, through the use of best available technologies. Apart from the economics, in many cases favourable, a number of barriers -

information, regulatory, institutional, financial-hinder the full uptake of these technologies. It is essential to try to remove such barriers to allow these emerging technologies to establish their operational reliability and cost competitiveness.

Every time energy-using capital stock and infrastructure is installed anywhere in the world, there is a unique opportunity to adopt climate-friendly technologies. If this opportunity is missed, not only will reductions in emissions not take place as early as they could, but potential suppliers of such technologies will have less incentive to develop them.

Early involvement of all actors concerned will help foster appropriate innovations and changes in long-term trends and infrastructures (e.g., in town planning) and to achieve emission reductions at minimum cost. This is all the more important in the context of electricity market liberalisation and deregulation of other energy activities, which may have lasting impacts on end-use energy prices, and uncertain effects on emission levels.

The time frame for reduction in emissions depends on the nature of the energy systems in each sector or country

The time frame to develop new technologies varies considerably depending on the type of existing equipment and underlying infrastructures, from district heating networks, gas-stations and pipelines, to electrified roads for public transport. Energy systems, from the production of energy to its end-use in specialised activities which contribute to economic development, are dynamic but with varying degrees of inertia. IEA countries have different economic structures and energy production and demand patterns, the result of different resource endowments, and choices made over decades. They will take these circumstances into account in the design of the most appropriate strategies for tackling their own GHG emissions.

COST-EFFECTIVE AND VIABLE ENERGY RESPONSES TO CLIMATE CHANGE

Preamble

While reading what follows, the reader should keep in mind the past trends of energy demand described earlier. They suggest that responses, as described below, should be assessed against their capacity to have a lasting influence on these trends, and their political acceptability.

Viable responses need to work with market dynamics

The promotion of environmentally sustainable economic growth requires providing and expanding energy services while simultaneously reducing their energy and CO_2 content. Market dynamics, the rigidity of infrastructures and attitudes, and the rate of capital stock turnover define the basic parameters of viable response options, pointing both to certain limitations and significant opportunities.

De-regulation of energy markets in IEA countries can be a vector for cost-effective responses

The current trend in energy market de-regulation in IEA countries should provide a more level playing field. This may not promote lower greenhouse gas emissions in all cases; yet it will provide a more cost-effective basis for actions aimed at reducing greenhouse gas emissions.

Market attitudes to risk shape the vital contribution of technology

Of particular importance are the perceptions of risk that may hinder new technology when it is regarded as compounding the intrinsic uncertainties of market competition. Yet dynamic implementation of the best available technology has a vital role to play in mitigation of the climate impact of the energy chain. A stable, transparent, and level playing field is essential to create clear market signals and to allow new climate-friendly technologies to be selected through normal commercial decision-making.

Policy also has a role to play in technological development. The major challenge is to combine a favourable context for research, development and deployment – both through public funding and an appropriate framework for private activity – with market-based evaluation and implementation mechanisms. The IEA / OECD Climate Technology Initiative (CTI) is exploring this challenge.

In a complex world, no single response option is uniquely viable

In theory, policies which act on prices (e.g., taxes on emissions) and those that act on quantities (e.g., regulations to limit emissions) can achieve equivalent emission reductions. In reality, however, their economic, political and social implications are likely to be very different.

As discussed in sections I and II, sectoral and national diversity, and the economic value of existing capital stock will heavily influence the impacts of such action. Because of widely varying effects of given policy instruments in different national contexts, it is difficult to define in general terms a mix of policies and measures which would be universally applicable. This suggests policies should be part of a learning process, whereby the implementation of responses provides information about their effectiveness and enables improved development of a flexible and adaptable policy framework over time.

Policies and measures can either directly target individual energy-users' behaviour or try to concentrate on a smaller number of institutional actors, government agencies, industrial users and the service sector, which influence individual consumers (e.g., construction companies which provide better insulated housing thanks to building codes). Experience shows that the latter group of actors may be easier to influence, even though in the long term, an effective climate change response will not be possible without support and action from the general public.

Potential policies include a variety of measures, such as: removal of various types of subsidies and price distortions; taxes and charges to raise energy prices; the accelerated introduction of reduced carbon and carbon-free technologies; voluntary agreements, standards and labelling programmes; target-related tradeable permits; and/or joint implementation.

Energy pricing which better reflects costs is potentially a win-win strategy

Privatisation and increased competition will tend to relate price-signals more closely to underlying costs and to introduce closer scrutiny of subsidies. In principle, full-cost pricing is a win-win long-term strategy since it provides an equitable basis for further policies to operate effectively within the market.

In practice, the situation is more complex because it is impossible to identify the "true" cost of all externalities, and an increase in price to reflect these externalities may not be painless in the short term. Nonetheless, moving towards elimination of subsidies and incorporation of identifiable externalities into prices will in general be a key component of climate change responses.

End-use prices are a strong determinant of energy consumption trends

Sectoral analysis indicates, in each end-use market, an inverse relationship between average energy prices and consumption per unit of GDP. This suggests that, even though short-run price-elasticities may be low, prices do influence consumers in selecting equipment and infrastructures, and in their use. Instruments affecting prices would provide a direct lever on consumption patterns, albeit with highly differentiated sectoral impacts. This explains why those IEA countries which have adopted carbon/energy taxation have partly exempted industry on competitiveness grounds.

Indeed, policies on energy prices have a variety of significant knock-on effects – notably on international competitiveness, aggregate welfare and income distribution – that need to be taken into account in their design and the evaluation of their practicality.

Climate-friendly technologies can be used in transport and electricity generation

A variety of options are available which could reduce the CO_2 impact of power generation for the same level of service provision: more efficient fossil technologies (e.g., cogeneration), renewable energies, large hydro dams and nuclear power stations. The drastic changes which have taken place in this area since 1973 indicate that the potential is large. Similarly, in the transportation sector, electrical, hybrid and hydrogen vehicles are available at varying levels of commercial viability.

Their applicability will vary depending on national contexts. In particular, their beneficial impact on CO_2 emissions may be offset by considerations of security of supply or other environmental issues. Furthermore, they have upstream impacts (e.g., dams for hydropower, long-haul natural gas shipping for power generation) that may affect their net GHG balance.

Furthermore, longer-term R&D is needed to develop cleaner and more efficient technologies for energy production and end-use

Cleaner and more efficient technologies can be introduced for energy production and end-use. However, deployment of new technologies is dependent on a mix of market "pull" and technological "push" (whether this triggers incremental improvements or technological breakthroughs). Ongoing technological progress requires commitments to long-term investments in research and development, particularly to achieve further decarbonisation in the power sector (e.g., with renewables or other non-fossil sources), to promote CO_2 capture and storage, to improve the efficiency of fossil fuel use, and to increase end-use energy efficiency.

Voluntary agreements, audits, labels and standards can address certain forms of market failure

When energy costs are a major input to an economic activity, energy consumption will generally be addressed in the most efficient way by private operators.

On the other hand, when energy costs represent a small input – in non-energy-intensive industries, services or households – operators may neglect profitable energy efficiency measures because they tend to focus on other, more important inputs. In such circumstances, audits together with voluntary agreements, standards, or labelling and information campaigns supported by energy service companies, may raise operators' concerns for the energy implications of their day-to-day operations and investment choices.

There is promising experience of such action, thanks to initiatives of a number of IEA governments promoting cooperation between public or private institutions and industry, services, and the household sector.

Efficiency standards have demonstrated their potential effectiveness

Trends in energy demand and related CO_2 emissions for the three services have shown that improvements in the energy efficiency of specific equipment have been more or less offset by a combination of more and/or larger equipment, as well as increase in use, largely as a result of economic growth.

Energy efficiency standards have been implemented for a range of end-uses (vehicles, building insulation codes, electric appliances, etc.). By accelerating the introduction of more efficient equipment, efficiency standards have had positive effects on both end-users' welfare and energy savings. Past trends suggest that further energy savings can be achieved by limiting the so-called rebound effect (i.e., the increase in use as a result of lower energy costs, which reduces the expected energy savings).

An interesting experience can be found in the US corporate average fuel economy standards for cars and light trucks which produced an overall reduction of 20% in fuel demand for transportation from past trends, mainly attributable to reductions in vehicle weight. Such measures may not be as effective in other countries, where there is much room for an increase in car numbers and use. However, such measures could be employed to promote non-conventional vehicles (natural gas, electric, hybrid...).

An equivalent tool could be a system of feebates / rebates providing a signal on the aggregate level of GHG emissions for each vehicle over their expected lifetime, which would gear consumer choices towards more climate-friendly technologies.

Tradeable permits offer flexibility in mitigation

Tradeable permits have demonstrated their effectiveness in achieving innovative and flexible cost-effective compliance with respect to atmospheric pollutants such as SO_2 and lead in gasoline.

International or national emissions trading offers a potentially efficient means of reducing the marginal and overall cost of controlling greenhouse gas emissions, while providing an incentive for early action by those participants for which it is cost-effective to do so. A further advantage of emissions trading is the incentive it would provide for investment and technology diffusion.

Applying such instruments to CO_2 emissions is attractive in principle, but many important issues have to be resolved: equity concerns in the allocation of emission quotas, potential transaction costs, and the necessity of an international monitoring system.

Joint implementation of projects and activities could complement an emissions trading scheme

Among instruments which either add to or may help to introduce a system of tradeable permits, joint implementation among Annex I countries or Activities Implemented Jointly (AIJ) between Annex I and non-Annex I countries could provide an option, subject to addressing issues such as baselines, performance monitoring, and crediting of avoided emissions. Furthermore, this offers helpful opportunities for cooperation with the developing world, in a way that is consistent with their current energy needs.

Many areas could be envisaged for Activities Implemented Jointly, such as methane leakage in natural gas pipelines, demonstration and deployment of renewable energy, and power-plant efficiency improvements.

Combining such potential instruments over three energy services is a complex task ...

Although some options are clearly geared specifically towards certain sectors (industry, utilities, services, households...), there are a number of possible combinations leading to different policy packages for the three energy services (transport, electricity, other stationary uses).

Cost-effectiveness is a key objective, but is highly dependent on national and sectoral situations. What is cost-effective in certain national circumstances may not be in others: experience is not necessarily transferable.

... requiring cooperation among IEA countries ...

Cooperation among countries to share their experiences and attitudes will, within the limits imposed by differing national circumstances, bring benefits ranging from replication of actions that have proved successful in one country to actions undertaken as a genuine common policy that might otherwise not be possible at national level.

Policies and measures might include agreement to act together in defined areas, leaving flexibility on how precisely to act; and measures and strategies to minimise the economic consequences of action, e.g., through changes in competitiveness.

... between IEA and non-IEA countries ...

These options are open to developing countries as well as IEA member countries, although, as with IEA member countries, their applicability depends a great deal on each country's economic, social and political circumstances and technical capabilities. Through the Climate Technology Initiative, as well as through other multilateral or bilateral means, IEA member countries are eager to share their experiences in these areas with developing countries.

Yet, the growing share of non-IEA countries in CO_2 emissions implies that, whenever possible, actions should be taken together. In particular, rapid infrastructure growth in developing countries provides important opportunities for cooperative action, including technology and voluntary agreements, labels and standards, and AIJ.

... and within each IEA country

Last but not least, within each individual country, policy packages to tackle the threat of global climate change will need strong cooperation between ministries (environment, energy, transportation, industry, finance, foreign affairs), businesses and the general public.

2

CO_2 EMISSIONS IMPLICATIONS OF THE IEA 1996 "WORLD ENERGY OUTLOOK" CAPACITY CONSTRAINTS CASE

INTRODUCTION

This chapter presents the implications for energy-related CO_2 emissions of the energy demand projections to 2010 contained in the Capacity Constraints Case developed by the IEA in 1996. This case provides a detailed assessment of likely energy market developments at a global and regional level, derived from the application of the large-scale World Energy Model. The IEA *World Energy Outlook* (WEO)* projections were prepared using the conventional approach to sectoral disaggregation in energy statistics, namely:

■ Primary energy supply;

■ Energy transformation, including refining and electricity and heat production;

■ Final consumption of individual fuels in industry, transport, commercial services and the residential sector.

The results of the WEO have been re-formatted and the main factors affecting them re-presented within the framework set out in the previous chapter. The results of the Energy Savings Case, the other case analysed in the 1996 WEO, are presented in Chapter 3 of this book.

1996 *World Energy Outlook*: Key Assumptions

Projections of energy demand to 2010 in the IEA *World Energy Outlook* were generated by the application of a large-scale econometric model, the World Energy Model. The model comprises a set of relationships involving energy demand, supply and price data and using the conventional approach to sectoral disaggregation in energy statistics.

Two cases are presented: the *Capacity Constraints Case* and the *Energy Savings Case*.

■ The most important assumptions behind the projections concern *economic growth*. These assumptions are the same for both baseline cases. GDP in OECD countries is assumed to grow by 2.5% per annum from 1993 to 2010, while world GDP is assumed to grow by 3.2%.

---

* *World Energy Outlook* (IEA/OECD, Paris, 1996).

- Assumptions concerning *energy prices* differ in the two cases:

  • In the Capacity Constraints Case, the average crude oil price is assumed to remain broadly flat at around $17/barrel (in real 1993 dollars) until 2000, then rise to $25/barrel in 2005 and remain at that level to 2010.

  • In the Energy Savings Case, the oil price is assumed to remain at $17/barrel over the entire forecast period.

- The other key assumption concerns *energy efficiency improvements*:

  • In the Capacity Constraints Case, no new policies are assumed, and energy efficiency improvements continue at their historical rates.

  • The Energy Savings Case assumes higher energy savings but does not specify the behavioural changes as well as government regulations and other non-price related measures needed to achieve the savings (see Chapter 3 for details).

Full details are provided in the IEA *World Energy Outlook* 1996 Edition, IEA/OECD Paris, 1996.

1996 *World Energy Outlook*: Key Results of the Capacity Constraints Case

- World primary energy demand is projected to be 46% higher in 2010 than in 1993. The average annual growth rate is projected to be 2.2%, the same growth rate as from 1971 to 1993.

- This increase in demand is expected to be met almost entirely by fossil fuels, which are projected to account for almost 90% of total primary energy demand in 2010.

- World oil demand is projected to rise from 70 million barrels per day (mb/d) in 1995 to 97 mb/d in 2010.

- The fuel mix in primary demand is not projected to change significantly: oil remains the main fuel, with 40% of demand in 2010, followed by coal, with 25%, and gas, the share of which increases by two percentage points to 24%. The share of nuclear power is projected to decrease from 7% in 1993 to 6% in 2010.

- A structural shift in the shares of different regions in world energy demand is projected to occur. The OECD share of world energy demand falls from around 55% in 1993 to less than 50% in 2010. The share of the developing countries in world demand for primary energy is projected to increase from 28% to almost 40% over the same period, due to rapid economic growth and industrial expansion, high population growth and urbanisation, and the replacement of traditional non-commercial fuels by commercial energy.

THE OUTLOOK FOR ENERGY DEMAND: AN ENERGY-RELATED SERVICES ANALYSIS

Electricity demand continues to increase in line with real income

In the OECD, fuel inputs to power generation are projected to continue to rise as electricity demand rises with increasing GDP. This is broadly in line with the nearly linear historical trend since 1971 (see Figure 2.1). There is a slight flattening of the curve from around 2000, as a result of several factors, notably:

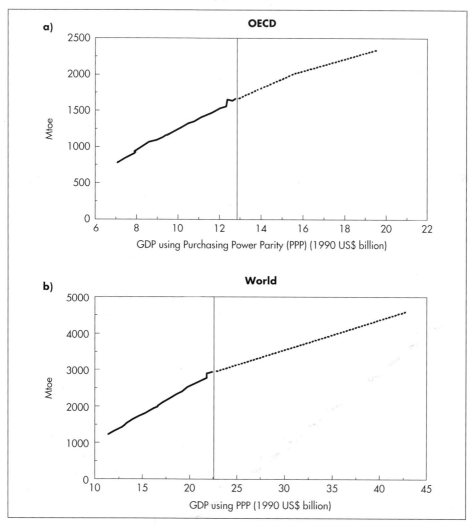

Figure 2.1
Fuel Inputs to Power Generation in the Capacity Constraints Case, 1971-2010

■ Increasing thermal efficiency in power generation, resulting mainly from the commissioning of high-efficiency natural-gas-fired Combined Cycle Gas Turbine plants (CCGTs).

■ The assumption of saturation of electrical equipment in final electricity uses in the household and, to some extent, the services sector. (This is partly offset by the rising number of households and the emergence of new electrical appliances and other equipment.)

For the world as a whole, there is a slightly more pronounced flattening of the curve of demand for fuel inputs against real GDP. This is mainly due to expected improvements in thermal efficiency — particularly in the countries of the former Soviet Union and Eastern and Central Europe (FSU/ECE).

Transport Sector demand grows rapidly, particularly outside the OECD

Transport sector demand for fossil fuels in the OECD rises to the early 2000s, broadly in line with the trend established since the second oil price shock (see Figure 2.2). The impact of rising GDP on demand diminishes slightly towards 2010, in response to three main factors:

■ An assumed increase in transport fuel prices due to increasing world oil prices.

■ Constraints on road traffic, including congestion, limitations on infrastructure development and environmental concerns (particularly local pollution problems).

■ Changes in the structure of GDP away from heavy industry towards services and a shift towards lighter materials (dematerialisation) that require fewer additional tonne-kilometres.

Demand in the world transport sector increases in a more linear fashion in relation to GDP because of a steady increase in non-OECD demand. This is mainly driven by rising demand for mobility in the developing countries and in the FSU/ECE, induced by increases in per capita income levels. (There is currently a low level of vehicle ownership in these regions.)

Growth in global demand for non-electrical energy in stationary uses slows after 2000

As can be seen in Figure 2.3, demand for energy in the non-electricity stationary uses sector in OECD countries is projected to remain relatively insensitive to GDP. In fact, demand has increased very little over the past 15 years in spite of an increase in real GDP of more than one third.

Figure 2.2
Transport Sector Demand in the Capacity Constraints Case, 1971-2010

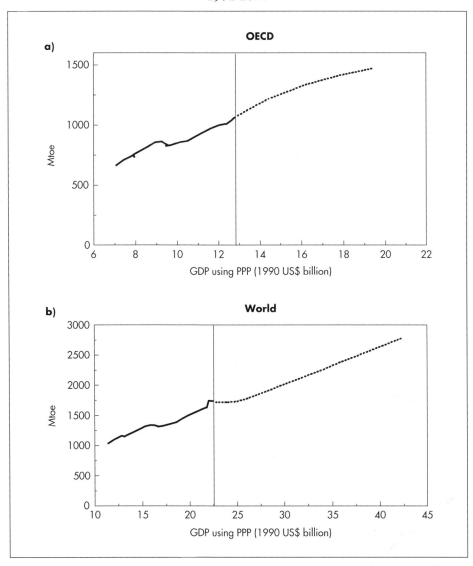

Demand is projected to increase very slightly to the end of the 1990s, partly because of historically low fossil fuel prices and continued modest growth in industrial output. (The troughs in energy demand since the second oil shock are closely related to periods of declining GDP and lower industrial output.) Higher prices over the period 2000-2005, coupled with the progressive saturation in demand for fossil fuels for household heating, result in stagnating demand to 2010.

Figure 2.3
Stationary Uses Sector Demand in the Capacity Constraints Case, 1971-2010

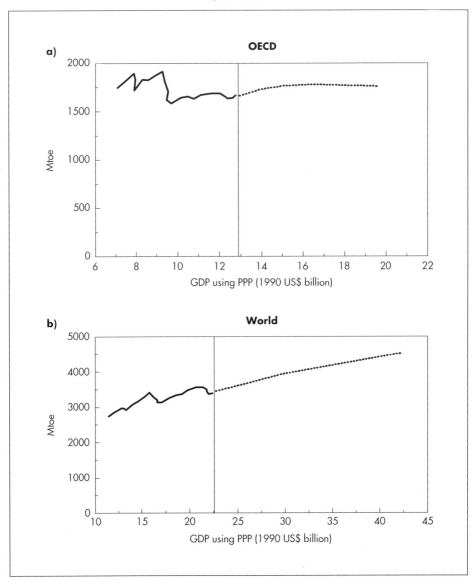

In non-OECD countries, demand for energy in stationary uses is projected to rise in line with GDP in a nearly linear manner. Demand would rise in the industrial, services and household sectors in response to rising incomes and rapid growth in manufacturing output, outstripping improved energy efficiency in industry. Relocation of heavy industries to non-OECD regions would continue to play an important role. For the world as a whole, primary energy demand in relation to GDP levels off slightly, reflecting stagnant demand growth in the OECD.

Figure 2.4
Total World Primary Energy Demand in the Capacity Constraints Case, 1971-2010

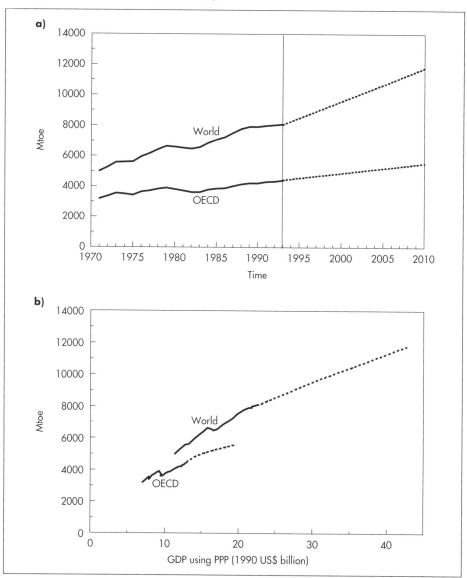

Aggregated energy demand in the three energy-related services for the OECD and the World is given in Figure 2.4. As can be seen from these figures, the aggregate demand projections for OECD are similar to past trends with respect to both time and real GDP. The projections for aggregate demand in comparison with GDP in energy-related services for the world as a whole, however, are slightly lower than past trends. This is mainly due to the expected slow growth in energy demand in the FSU and efficiency improvements in electricity generation. Global trends excluding the FSU are barely changed.

THE OUTLOOK FOR CO_2 EMISSIONS

Total energy-related emissions continue to rise steadily

By 2010, global CO_2 emissions are projected to have grown by almost 50% from their 1990 level under the Capacity Constraints Case, from 21.1 billion tonnes in 1990 to about 31.5 billion tonnes. The bulk of this increase is expected to occur in non-OECD countries, where CO_2 emissions are projected to more than double. In the OECD countries, CO_2 emissions in 2010 are likely to exceed their 1990 level by close to 30%.

In Figure 2.5, prospects for CO_2 emissions are plotted for three regions and for the world as a whole. In FSU/ECE, improvements in energy efficiency in all sectors are expected to have a greater effect than increases in energy purchases, as incomes rise back to 1990 levels by 2010. CO_2 emissions in FSU/ECE are projected to be slightly lower in 2010 than in 1990. CO_2 emissions in OECD countries grow largely in line with past trends. In the rest of the world, the increased use of gas slows the upward trend. Gas nevertheless continues to account for most of the growth in CO_2 emissions. For the world as a whole, the net effect is a slight reduction on the past rate of growth.

Per capita emissions differences between regions remain large

It is important to note that by 2010 there will still exist substantial differences in per capita CO_2 emissions across countries, although there will be some convergence between the OECD and non-OECD regions. Because population will be increasing rapidly, the overall growth rate in CO_2 emissions in developing countries will be more than double the growth rate for per capita CO_2 emissions.

Electricity generation and transport account for most of the emissions increase

The breakdown of total world CO_2 emissions among energy services is similar to that for energy consumption in these services. At a global level, energy demand for power generation and transport is projected to increase by almost 3% per annum, whereas growth in stationary use is expected to be only 1.6% per annum. This explains the significant drop in the relative contribution of stationary energy use to total CO_2 emissions. However, it is expected that stationary sectors will still have the highest share of total world emissions by 2010 (see Figure 2.7).

This trend is even more apparent in the OECD countries. Almost no increase in CO_2 emissions is expected in the stationary sectors. However, the shares of fuels in total stationary use are projected to change significantly. As illustrated in Figure 2.8, natural gas is expected to increase its share and become the main contributor by 2010. This is almost totally due to the declining share of coal. Power generation-related emissions are projected to increase at the highest rate in the OECD,

Figure 2.5
Energy-Related CO_2 Emissions and Real GDP, 1971-2010

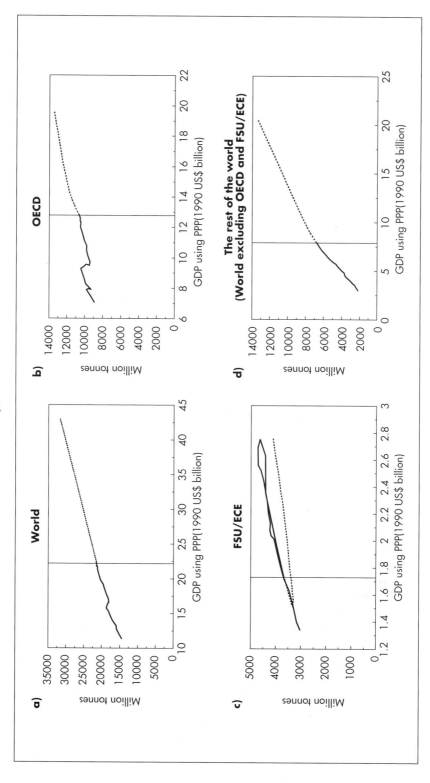

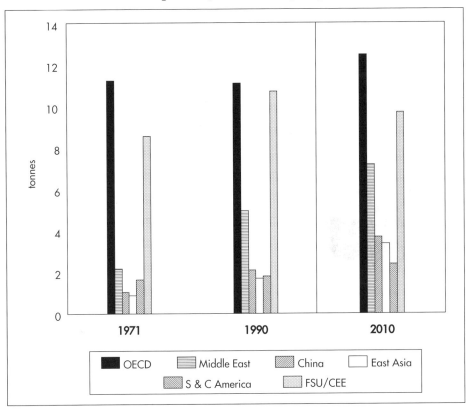

Figure 2.6
Per Capita CO_2 Emissions by Region

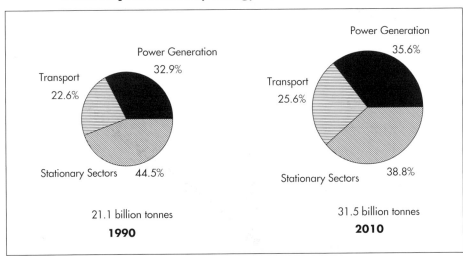

Figure 2.7
World CO_2 Emissions by Energy-Related Services Sectors

slightly over 2% per annum. It is expected that the share of gas in power generation will double over the outlook period while the shares of coal and oil will decline significantly. This reflects the considerable advantage of CCGTs over all other types of plant.

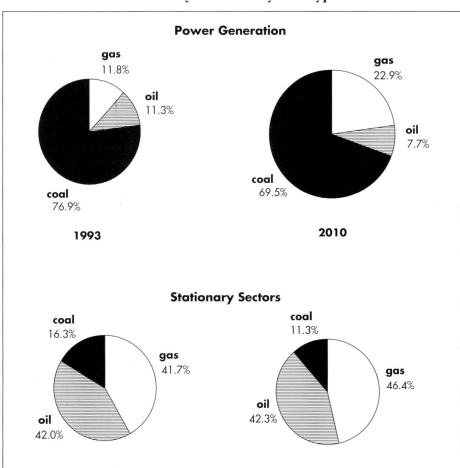

Figure 2.8
OECD CO_2 Emissions by Fuel Type

Coal is the major fuel for power generation and industry in two of the fastest growing non-OECD areas, South Asia and China, and coal is the most carbon-intensive of all conventional fossil fuels. In 1993, coal accounted for around 75% of total electricity generation in China, while in South Asia it accounted for 60%. Electricity demand in both of these areas is projected to increase by 6 to 7% per annum during the outlook period, and the share of solid fuels in electricity generation is likely to remain broadly unchanged. The power generation sector in these countries will contribute significantly to the projected increase in developing countries' CO_2 emissions, despite assumed improvements in the efficiency of power generation in all developing countries. Transportation is also expected to contribute significantly to the growth of CO_2 emissions from developing countries — 4.2% per annum between 1990 and 2010.

Transport in the FSU/ECE is expected to show the biggest increase in CO_2 emissions among energy-related service sectors. The cumulative increase is projected to be above 60% over the outlook period, implying an annual growth rate of close to 3% per annum. On the other hand, the overall projected emission levels for the power generation and stationary use sectors in 2010 are almost the same as the 1993 levels, despite a recovery after the economic collapse of the early 1990s. This is mainly due to expected efficiency improvements in these sectors.

SUMMARY

In the Capacity Constraints Case, world energy demand grows by 46% between 1993 and 2010. Over the same period, world CO_2 emissions grow at approximately the same rate, increasing by 50% between 1990 and 2010.

Comparisons between the projections and historical trends in energy consumption and CO_2 emissions for the key energy-related services identify the main factors producing deviations in trend in the Capacity Constraints Case as:

- Energy intensity continues to decrease.
- Possible saturation of electricity equipment in households and some service sectors may be partly offset by the emergence of new types of appliances.
- Increases in energy prices have a small effect in lowering energy demand.
- Possible constraints develop on growth in road transport (e.g. congestion).
- World industrial energy demand increases in the context of strong economic growth.

3

CO_2 EMISSIONS IMPLICATIONS OF THE IEA 1996 *"WORLD ENERGY OUTLOOK"* ENERGY SAVINGS CASE

This chapter describes the projections for energy demand and related CO_2 emissions in the Energy Savings Case of the 1996 edition of the IEA's *World Energy Outlook*; it makes comparisons with the Capacity Constraints Case.

THE ENERGY SAVINGS CASE ILLUSTRATES THE POTENTIAL FOR COST-EFFECTIVE ENERGY EFFICIENCY MEASURES

> **The Energy Savings Case**
>
> The Capacity Constraints Case (described in Chapter 2), assumes the continuation of past policies — including energy efficiency and energy conservation policies. As explained in this chapter, new policies are needed to go beyond the energy savings achieved in the Capacity Constraints Case.
>
> The *Energy Savings Case* assumes the attainment of such additional energy savings, but does not investigate or specify the policies needed to achieve them. It represents only an illustrative example of higher energy saving than the Capacity Constraints Case.

The Energy Savings Case incorporates assumptions of changes in the uses of technologies and in consumer behaviour that lead to lower energy intensity than historical trends suggest. It does not attempt to explain what policies would be required to bring about those changes. Key assumptions are given in the box on page 54.

Market and institutional distortions may prevent energy users from taking measures that would otherwise be cost-effective for them and reduce greenhouse gas emissions. The most economic commercial technology is not always used, and energy equipment is not always used as efficiently as it could be. Examples of why this happens include regulatory impediments to efficient fuel use (e.g. measures aimed at achieving non-energy related goals, such as improved road safety), price distortions and subsidies, inadequate information flows and imperfect capital markets that limit the amount of credit available to firms, families or institutions for investments. Other factors include low priority given by decision makers to energy savings.

The assumptions made in the Energy Savings Case result in a significant reduction in the rate of energy demand growth over the outlook period and reduce the need for additional energy-production capacity. Energy markets are assumed to expand in an orderly fashion to meet relatively modest growth in energy demand. Therefore upward pressure on prices does not arise: energy prices are assumed to remain broadly flat until 2010.

The Energy Savings Case thus provides an indication of how world energy demand and supply could be affected by a significant rise in the use of cost-effective energy intensity improvements. Cost-effective here means that the individual or organisation undertaking the improvement in energy intensity saves energy of a value that exceeds the costs of undertaking the improvement *without taking into account any environmental or social costs or benefits*. The Energy Savings Case includes no new technologies that are not cost-effective in the sense described above. However, major changes in the way consumers make their choices in the consumption of energy and other goods are assumed.

Key Technology Assumptions in the Energy Savings Case

Transport Sector

It is assumed in the Capacity Constraints Case that the energy intensity of new passenger cars in OECD countries decreases by 0.5% per annum in Europe and the Pacific and by 0.2% per annum in North America, consistent with historical trends. In the Energy Savings Case, these decreases are assumed to occur at a substantially higher rate, 1.5% annually, throughout the OECD. Similarly, in the Energy Savings Case, the share of diesel cars in new car registrations in Europe is assumed to rise from 18% in 1993 to 32% by 2010, compared with 24% in the Capacity Constraints Case. Given the better fuel economy of diesel engines, this also contributes to an improvement in overall European transport efficiency and to a lower increase in fuel use and CO_2 emissions. Reductions in the energy intensity of new trucks and aircraft are assumed to be twice the size of those in the Capacity Constraints Case. In the FSU/ECE and developing countries, additional intensity reductions are assumed of around 1% annually below those in the Capacity Constraints Case in all major transportation segments (passenger cars, trucks and air transportation). It is also assumed that vehicle efficiency gains are not offset by corresponding increases in vehicle size, population and kilometres driven, as has tended to be the case in the past.

Power Generation Sector

The thermal efficiency of new combined cycle gas turbine plants in the OECD is assumed to increase gradually and to approach 56% by 2010 in the Capacity Constraints Case and 60% in the Energy Savings Case. This contrasts with an average efficiency of 42% for the current combined cycle gas fired generation stock and 35% average efficiency for fossil-fuelled plant as a whole. In addition, the share of heat from combined production of heat and power (CHP) plants in total electricity and heat generation is assumed to be 6 percentage points higher in the Energy Savings Case than in the Capacity Constraints Case. This is an important assumption because most of this additional heat displaces fossil fuels in the stationary uses sector. In the FSU/ECE and the developing countries, higher thermal efficiency is also assumed for new power plants.

Stationary Fossil Fuel Uses

In this sector, it is not possible to make explicit assumptions on the contribution of specific technologies to energy intensity improvements because of the high level of aggregation. The improvements in energy use assumed in the Capacity Constraints Case reflect historical trends in energy intensity. In the Energy Savings Case, small additional improvements in energy intensity beyond the contribution of CHP noted above are assumed to occur in all three main regions.

Principal Differences Between the Energy Savings (ES) and The Capacity Constraints (CC) Cases

The Energy Savings Case projects lower energy demand and lower CO_2 emissions in 2010.

    932 mtoe or 8% lower energy demand than in the CC Case.

    2.7 billion tonnes of CO_2 emissions or 9% lower.

Global CO_2 emissions grow by 36.5% between 1990 and 2010 in the ES Case, compared with 49.3% in the Capacity Constraints Case.

The reductions in CO_2 emissions in billions of tonnes by sector are:

power generation	1.4	(15% lower than in the CC Case)
stationary fossil fuels	1.0	(8% lower)
transport	0.3	(5% lower)

The reductions in CO_2 emissions in billions of tonnes by world regions are:

OECD	1.5
FSU/ECE	0.5
Rest of World (ROW)	0.7
Total	2.7

Full details of these two cases may be found in *World Energy Outlook*, IEA/OECD, 1996.

Figure 3.1 plots global primary energy demand and energy-related CO_2 emissions over time, for the Energy Savings Case and for the Capacity Constraints Case. By 2010, the gap between the two cases amounts to 930 million tonnes of oil equivalent and 2.7 billion tonnes of CO_2, or about 8% of projected demand[1] and 9% of projected emissions in the Capacity Constraints Case.

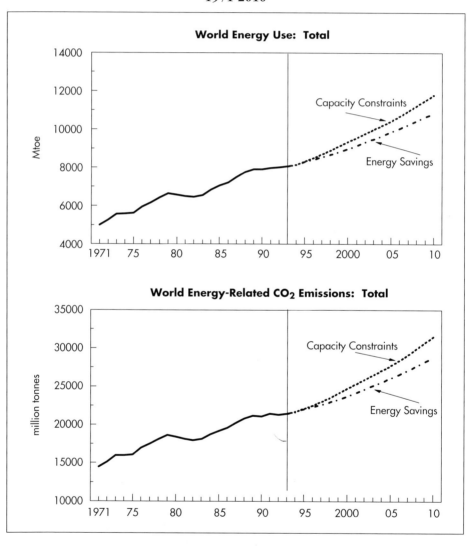

Figure 3.1
World Primary Energy Demand and Related CO_2 Emissions, 1971-2010

---

[1] The potential is slightly greater than that indicated in the Energy Savings Case because that case also assumes flat energy prices compared to rising energy prices in the Capacity Constraints Case. The Energy Savings Case also excludes the potential for cost-effective efficiency gains in China because of difficulties in estimating current energy efficiency levels and the scope for improved efficiency.

Figure 3.2 plots energy-related CO_2 emissions against real GDP for the period 1971 to 2010. In the Energy Savings Case, there is a slight decoupling of emissions from economic growth: emissions grow by around 34% (compared to 47% in the Capacity Constraints Case) against a near-doubling of GDP.

Figure 3.2
World Energy-Related CO_2 Emissions and Real GDP

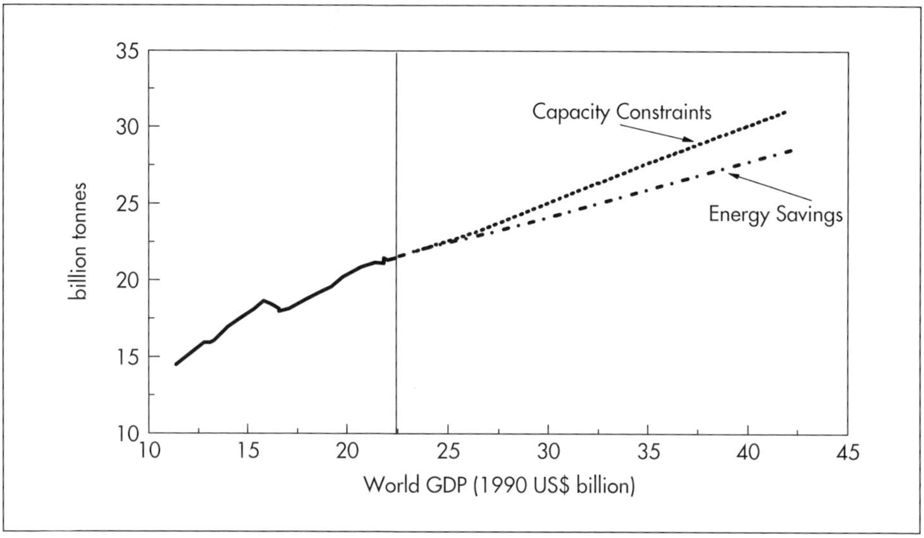

SECTORAL VARIATIONS

On a global scale, in the Energy Savings Case the greatest potential for energy efficiency improvements and corresponding emissions reductions — both in absolute terms and in proportion to current emission levels — is to be found in the electricity generation and stationary uses sectors (see Figure 3.3).

In electricity generation, the potential for CO_2 emissions reduction amounts to 1.4 billion tonnes by 2010, equivalent to almost 15% of the total emissions projected in the Capacity Constraints Case. These savings result from three factors:

- **Higher thermal efficiencies in generation.** The emissions gap between the two cases increases over time, due to lead times in introducing more efficient technology. This technology notably includes combined cycle plants fired by natural gas or oil, coal-fired plants and co-generation plants. Higher efficiency accounts for around two thirds of the gap between the Capacity Constraints and Energy Savings Cases for power generation.

Figure 3.3
**Energy-related CO_2 Emissions Reduction in Energy Savings Case Compared With Capacity Constraints Case, 2010
(Million tonnes, and %, by sector, of total emission saving)**

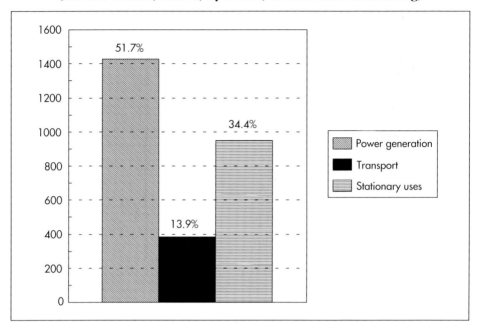

- **A lower level of carbon intensity of fuel inputs to generation**[2]. Additional capacity requirements are assumed to be met almost entirely by fossil fuels, and the contribution of hydroelectric and nuclear power in absolute terms is assumed to be the same in both cases. The share of fossil fuels in inputs to generation is thus lower in 2010 in the Energy Savings Case (see Figure 3.4) because total generation is less. This effect reduces the carbon intensity of fuel inputs to generation. The effect is modified by the fact that investment in new gas plants is reduced more than investment in new coal plants in the Energy Savings Case.

- **A slightly lower level of end-user electricity demand**, accounting for around 0.3 billion tonnes of CO_2 emissions or 3% less than foreseen in the Capacity Constraints Case.

In stationary fossil fuel uses, CO_2 emissions in the Energy Savings Case are just under 1 billion tonnes or 8% below the level projected in the Capacity Constraints Case. The residential/service sector accounts for around 55% of the total emissions reduction. The remaining 45% occurs in industry. Part of the

---

2   Carbon emissions per unit of heat content of the fuel input.

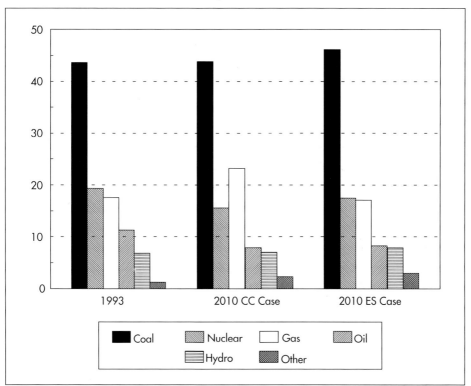

Figure 3.4
Fuel Inputs to Power Generation, 2010
(% of total inputs to generation)

overall reduction is explained by increased heat recovery in power generation. This heat substitutes for fossil fuels in stationary uses, resulting in lower emissions.

The overall energy efficiency gains and CO_2 emissions reductions in the transport sector are small, amounting to an estimated 300 million tonnes or 5% of emissions in the Capacity Constraints Case in 2010. The gains are achieved gradually over the period to 2010 as new, more efficient car and truck designs are introduced, assuming that efficiency gains are not offset by behavioural changes resulting in higher consumption.

REGIONAL VARIATIONS

The efficiency improvement effects are not evenly distributed among regions. Figure 3.5 shows that, in absolute terms, the OECD accounts for 54% of the CO_2 emissions difference between the two cases. The difference in the

Figure 3.5
**CO$_2$ Emissions Reduction in the Energy Savings Case
Compared with the Capacity Constraints Case, 2000 and 2010 by Region
(Million tonnes)**

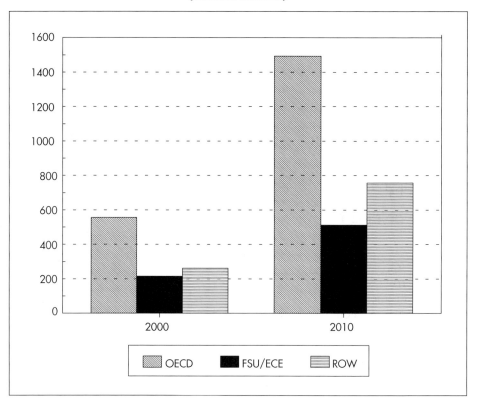

FSU/ECE reflects the considerable scope for energy efficiency improvements through the transfer of technology as these countries continue to implement market reforms.

In the OECD, the largest CO$_2$ emissions reductions are in the power generation sector (see Figure 3.6). More efficient generation technology than in the Capacity Constraints Case is introduced gradually from the mid-1990s through 2010. Final demand also increases more slowly, reducing the need for new capacity which is primarily fossil fuel based. In the FSU/ECE, emissions reductions are shared more evenly between the power and stationary uses sectors: there is potential for energy efficiency gains in end-use applications in the industrial, service and residential sectors. Based on the observed impact of two oil shocks on OECD countries, part of these gains will remain once economic growth resumes. In the developing countries, there is a more even distribution of the potential across the three main sectors. The sectoral and regional shares in total emissions reductions are set out in Table 3.1.

Table 3.1
Analysis of the Reductions in Energy-Related CO_2 Emissions in the Energy Savings Case Compared with the Capacity Constraints Case, 2010
(% shares of total reduction by region and sector)

Sector	OECD	FSU/ECE	ROW	World
Power generation	34.5	8.2	9.0	51.7
Transport	7.6	0.6	5.7	13.9
Stationary uses	11.8	9.9	12.7	34.4
Total*	54.0	18.6	27.4	100

* Excluding international bunkers.

HOW ENERGY SAVINGS MIGHT BE ACHIEVED IN THE ENERGY SAVINGS CASE

It is important to consider carefully the key assumptions concerning energy intensity improvements in the Energy Savings and Capacity Constraints Cases. These improvements are defined in the *World Energy Outlook* as those resulting from government actions (such as improved information, standards and regulations), from technical progress or from changes in consumer attitudes and behaviour. These are termed non-price-related energy intensity improvements. The Capacity Constraints Case assumes that autonomous energy intensity improvements will follow the historical trend. The Energy Savings Case takes a more optimistic view. It includes effects that are feasible in the sense that no new technologies are involved and the individual improvements assumed have already been shown to be cost effective (i.e. zero or negative net cost as calculated using a conventional discounted cash flow calculation).

A fundamental question arising from this analysis is whether non-price-related energy savings of the scale indicated in the Energy Savings Case are likely to occur without the introduction of additional government policies. This issue is considered in the next chapter.

SUMMARY

The Energy Savings Case in the 1996 *World Energy Outlook* incorporates an assumed potential for cost-effective reductions in energy use and related CO_2 emissions amounting to at least an 8% cut in world energy demand and a 9% reduction in emissions. This is a clear departure from historical trend. The power sector is identified as the single largest source of emissions reductions, accounting for around half of the world total, followed by stationary uses. The OECD accounts for more than 50% of the total emissions reduction.

This potential, which is assumed to be realised in the Energy Savings Case but not in the Capacity Constraints Case, is explained by the existence of barriers to energy-efficient choices: for example, lack of information, strong preferences for alternative expenditures, regulatory impediments, and limitations on income and credit. Active government policies and measures could assist in overcoming these barriers and in realising at least part of this potential.

4

REALISING ENERGY SAVINGS POTENTIAL TO 2010: THE POLICY RESPONSE

INTRODUCTION

The 1996 *World Energy Outlook* intentionally avoided discussion of what combination of policies might bring about the technological and behavioural changes that produce the reductions in energy intensity assumed in the Energy Savings Case. The analysis below is intended to provide insights into how those reductions might be realised through policy action by governments, taking account of cost-effectiveness and political feasibility. Both price-related and non-price-related policies are considered here.

Types of Energy-Saving Policies

Government policy can influence the behaviour of energy users, their decisions and technological development. The development of policy is complex, involving feedback, experimentation and learning-by-doing. The sequence is as follows:

i.   governments agree to certain targets, e.g. to reduce CO_2 emissions;

ii. governments then adopt policies that require, persuade or encourage energy producers and consumers to take actions that will reduce CO_2 emissions;

iii. consumers respond to these policies by changing their behaviour or modifying their lifestyles, producers by changing management rules or taking technical measures that reduce CO_2 emissions.

Section III of Chapter 1 - the IEA Statement on the Energy Dimension of Climate Change - lists different types of policy that governments can choose to reduce CO_2 emissions. These can be split into:

Non-price policies:

■ Information policies, such as energy labelling, partial or full financing of energy audits, creation of energy centres that provide energy-saving advice and information, "best practice" schemes (that finance demonstration projects and disseminate the results) and advice on technology procurement;

- Regulations that limit consumers' choices or require changes in their behaviour, including building regulations, energy-efficiency standards and speed limits;

- Voluntary agreements with industrial and commercial establishments to meet agreed targets.

Price policies:

- Policies that directly alter the end-use prices of fuels, e.g. removal of subsidies for energy production;

- Policies that include in prices a component that represents the environmental cost of the CO_2 emissions (through a carbon tax or tradeable CO_2 emission certificate).

Cost-Effective Energy-Saving Policies

Some of the actions taken by energy end-users to save energy are "no-regret" actions. That is, they show a net cash benefit without taking account of any environmental benefits. (Some authors include environmental benefits unrelated to the greenhouse effect along with cash benefits in this definition of no-regret actions.) The adoption of cost-effective, no-regret actions has considerable potential. The barriers to them were described in Chapter 3. More details on economic, cost-effective energy saving potential are given in the box below.

All IEA Members adopted energy-saving policies after the oil price shock of 1973-74. These policies remain an important component of their energy policies. Improved energy efficiency through strong government efforts is included in the IEA's Shared Goals. The mix of policies varies among countries depending on their circumstances, but to date the accent has been primarily on information policies.

Information policies can encourage the adoption of cost-effective energy saving and CO_2 abatement policies by showing that they are in the interests of individual energy consumers. Consumers can save money by saving energy. Such policies cost little and are non-controversial.

How much can energy consumption and CO_2 emissions be reduced by relying on information policies alone? Will it be necessary for governments to consider tougher regulations or the inclusion of a carbon tax in fuel prices to reduce CO_2 emissions below their 1990 levels by 2010? Some countries (Norway, Sweden, Denmark, the Netherlands and Finland) have already introduced carbon taxes. The United States uses emissions trading schemes for SO_2 emissions from power plants.

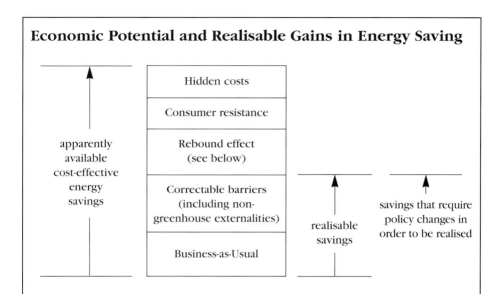

The full height of the above column represents the total of apparently available cost-effective energy savings. The column is split into segments, some of which represent savings that are realisable in response to appropriate government policies and some of which are not. Hidden costs are those that the energy consumer pays but which are not readily apparent to the analyst. Consumer resistance means that consumers are unresponsive to policies. The "rebound effect" is the additional use of energy that follows price reductions and corresponding increase in real income when efficiency of use increases. "Realisable" energy savings are those undertaken under Business-as-Usual circumstances together with savings that can be brought about by government action to remove barriers.

Adapted from "The Costs of Limiting Fossil Fuel CO_2 Emissions: A Survey and Analysis, M. Grubb, J. Edmonds, P. teh Brink and M. Morrison, *Ann. Rev. Energy Econ.* 1993, 18: 397-478.

Estimates of the technical potential for energy saving vary widely (irrespective of whether the saving is cost-effective). Grubb et al[1] quote two different studies for the United States. The Rocky Mountain Institute estimates that at additional costs of up to 4 cents per kWh, up to 70% of total electricity consumption could be saved. A study by the Electric Power Research Institute (EPRI) estimates that at additional costs of up to 13 cents per kWh, up to 30% of total electricity consumption could be saved. These dramatic differences arise from alternative assumptions regarding the sequence in which different technologies are introduced, the discount rates

---

[1] "The Costs of Limiting Fossil Fuel CO_2 Emissions: A Survey and Analysis," M. Grubb, J. Edmonds, P. ten Brink and M. Morrison, *Ann. Rev. Energy Econ.* 1993.

used, capital and operating costs of energy-saving technologies, prevailing energy prices, etc. Estimating what cost-effective actions are actually achievable is even more difficult. Among other factors, the responses of energy consumers to such policies must be assessed.

After considering a range of estimates, Grubb et al suggest:

> "Over a couple of decades, targeted energy-efficiency programmes might reduce energy demand by up to 20% of the level projected in the absence of any such policies, at costs lower than that of the displaced supply."

This implies an average rate of decline in energy intensity of 0.9% per annum (measured in terms of Final Energy Demand per unit of GDP).

The scope for cost-effective energy savings and the policies necessary to achieve them are discussed in the Contribution of Working Group III to the 1995 IPCC Second Assessment Report. The Report concludes that

> "Despite significant differences in views, there is agreement that energy efficiency gains of perhaps 10 to 30% above baseline trends over the next two to three decades can be realized at negative to zero net cost (negative net cost means an economic benefit). With longer time horizons, which allow a more complete turnover of capital stocks, and which give research and development and market transformation policies a chance to impact multiple replacement cycles, this potential is much higher. The magnitude of such "no regret" potentials depends on the existence of substantial market or institutional imperfections that prevent cost-effective emission reduction measures from occurring. The key question is then the extent to which such imperfections and barriers can be removed cost-effectively by policy initiatives such as efficiency standards, incentives, removal of subsidies, information programmes, and funding of technology transfer.
>
> Progress has been made in a number of countries in cost-effectively reducing imperfections and institutional barriers in markets through policy instruments based on voluntary standards, and energy efficiency incentives, product efficiency standards, and energy efficiency procurement programmes involving manufacturers, as well as utility regulatory reforms. Where empirical evaluations have been made, many have found the benefit-cost ratio of increasing energy efficiency to be favourable, suggesting the practical feasibility of realizing "no-regret" potentials at negative net cost. More information is needed on similar and improved programmes in a wider range of countries."

The second IEA Modelling Seminar, titled "Closing the Efficiency Gap in Energy Responses to Climate Change - Potential for Cost-effective Energy and Carbon Efficiency Improvements," held in Paris on 20-21 November, 1996, confirmed the above approach. It defined the economic potential for reducing greenhouse gas emissions as "energy efficiency improvements that could be achieved cost-effectively in the absence of market barriers." This seminar emphasized the importance of government action in achieving cost-effective emission reductions. It also emphasized that some of the policy measures would encounter considerable political difficulties in many countries.

Figure 4.1 plots annual energy intensities for OECD regions from 1960 to 1995. Table 4.1 lists annual energy intensity changes from 1961 to 1995. Energy intensities were uneven, but rose slightly from 1960 to 1974, suffered sharp falls in 1974 and 1975 as a result of the first oil shock and the resulting drop in economic activity, then rebounded slightly in 1976. A further sharp fall occurred in 1980 and 1981, following the oil price rise of 1979. Thereafter, energy intensities fell steadily, flattening out from 1986 onwards. Figures 1.3 and 1.4 in Chapter 1 show that the principal response to rising oil prices was a reduction in the consumption of fossil fuels for stationary end-uses.

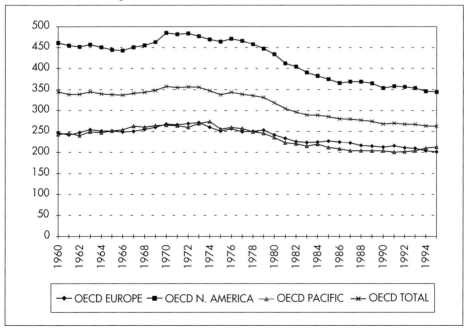

Figure 4.1
Energy Intensities in OECD Regions
TPES*/GDP (toe/m US$ 1990 prices and ppps)

* Total Primary Energy Supply.

The figures in Table 4.1 need to be considered together with average annual changes in energy intensity from 1993 to 2010 and the percentage changes in CO_2 emissions from 1990 to 2010 projected in the two WEO cases given in Table 4.2. By 2010, CO_2 emissions in OECD countries would be 27.4% above 1990 levels in the Capacity Constraints Case. To reduce this level to zero using energy efficiency measures alone would require an annual average decrease of 1.9%. To reduce emissions in 2010 to 10% below 1990 levels would require annual average decreases in energy intensity comparable to those following the oil price rises of 1973-74 and 1979. The same requirements apply to each of the three OECD regions: Europe, North America and the Pacific.

Table 4.1
Energy Intensity: Percentage Change from Previous Year
TPES/GDP (US$ 1990 prices and ppps)

	OECD Total	OECD N. America	OECD Pacific	OECD Europe
1961	- 1.84	- 1.35	0.97	- 1.86
1962	0.14	- 0.69	- 2.18	2.31
1963	1.70	1.15	3.78	2.80
1964	- 1.46	- 1.40	- 0.90	- 1.30
1965	- 0.47	- 1.28	1.65	0.05
1966	- 0.36	- 0.43	1.09	- 0.99
1967	1.31	1.81	3.48	0.56
1968	0.76	0.97	- 1.00	1.87
1969	1.19	1.79	1.48	1.89
1970	2.82	4.76	0.46	2.95
1971	- 0.72	- 0.71	- 0.65	- 0.45
1972	0.33	0.43	- 1.27	0.93
1973	- 0.20	- 1.36	3.38	0.97
1974	- 2.31	- 1.60	1.80	- 4.13
1975	- 2.76	- 1.15	- 6.61	- 3.59
1976	1.81	1.46	1.54	2.26
1977	- 1.42	- 1.07	- 1.03	- 2.79
1978	- 1.00	- 1.74	- 3.02	0.38
1979	- 1.28	- 2.31	- 1.48	1.13
1980	- 3.86	- 3.00	- 4.04	- 4.42
1981	- 4.29	- 4.94	- 5.15	- 3.35
1982	- 2.83	- 1.91	- 1.23	- 3.35
1983	- 2.29	- 3.41	2.54	0.98
1984	- 0.28	- 2.11	2.30	0.41
1985	- 1.17	- 2.03	- 3.61	1.16
1986	- 1.90	- 2.52	- 1.55	- 0.99
1987	- 0.18	0.94	- 2.22	- 1.06

1988	- 0.98	0.04	0.08	- 2.68
1989	- 0.95	1.01	0.23	- 0.73
1990	- 2.30	- 3.09	0.13	- 1.10
1991	0.64	1.22	- 1.53	1.52
1992	- 0.62	- 0.42	0.57	- 1.83
1993	0.11	- 0.85	0.97	0.12
1994	- 1.14	- 2.13	2.95	- 2.00
1995	- 0.09	- 0.40	1.49	- 0.17

Table 4.2
World Energy Outlook 1996

	OECD Total	OECD N. America	OECD Pacific	OECD Europe
Average Annual Percentage Changes in Energy Intensity				
1993-2010				
ES Case	- 1.5	- 1.6	- 1.1	- 1.8
CC Case	- 1.1	- 1.2	- 0.7	- 1.3
Total Percentage Change in CO_2 Emissions				
1990-2010				
ES Case	13.4	16.8	24.2	3.4
CC Case	27.4	29.9	43.1	17.1

The situation will vary substantially among countries because:

a) their energy consumption patterns and circumstances differ (industrial and economic structure, relative prices, etc.);

b) their current levels of energy intensity differ;

c) their vigour in pursuing energy efficiency policies affecting final demand differs (as opposed to policies affecting energy transformation).

This analysis suggests that much more effective and onerous energy efficiency policies than have been used so far will be necessary to achieve ambitious reductions in CO_2 emissions by 2010. Clearly, fuel switching and the use of more carbon-free fuels can also contribute to reducing CO_2 emissions. In increasingly liberalised and competitive energy markets, policies that increase the prices of CO_2 emitting fuels relative to other fuels and products can provide direct price signals for greater energy saving and relative price signals for fuel switching.

Political acceptability: What might be most economic may not be achievable

Notionally, policy actions for each country or region can be ranked according to their cost-effectiveness (not taking environmental or social costs into account), starting with low-cost measures (including negative net cost or "win-win" measures), followed by moderate and high-cost measures. Although some "no-regret" energy savings to reduce CO_2 emissions can be realised at no net cost to the consumer, a low cost would be involved at national level to introduce policies to overcome the current barriers. Once these savings have been implemented, market-based policies would provide a least-cost mechanism for achieving further energy savings, provided market forces are operating and prices adjust quickly. A carbon tax or system of tradeable permits would increase the cost of fossil fuel use to consumers. The aim of these policies would be to induce energy producers and consumers to invest in technology, product design or other means to reduce their CO_2 emissions. They would thus avoid paying the carbon tax or purchasing emission certificates. These avoidance costs would increase with the level of carbon tax or the limitation on carbon emissions imposed by the emission certificates. In addition, some national costs would be incurred in administering these schemes. Thus, although market-based policies would be least-cost measures to meet a specified reduction in CO_2 emissions, their cost would increase with the size of that reduction.

Standards and regulations are generally higher-cost measures than market-based policies for a given level of energy saving or CO_2 emission reduction. Standards and regulations may be necessary where market forces are not fully effective. However, only in a perfect world could policy makers use these approaches in rising order of cost, because of political constraints, imperfect information and uncertainty. In some cases, the energy saving achieved by a particular policy will depend on what other policies are already in place — i.e. the order of introduction of policies is an important factor.

In many countries, the least-cost strategies for reducing emissions, including no-net-cost measures, may be politically difficult. For example, significantly higher taxes on energy, balanced by tax reductions elsewhere in the economy, or removal of subsidies, have small economic cost or zero cost in certain cases.[2] However, the imposition of additional gasoline taxes in the United States or the removal of subsidies for coal production in Germany would cause severe political difficulties for the governments concerned. Political difficulties may be even greater in developing countries, where it is harder to demonstrate benefits and where such policies may have seriously adverse implications for the poorest sections of society. Moreover, the pressure from electorates to act on climate change is generally lower. Tax restructuring that raises the cost of energy, even if it reduces other prices, can be extremely difficult politically,

---

2   "A Green Scenario for the UK Economy," T. Barker and R. Lewney, in *Green Futures for Economic Growth*, edit. T. Barker, Cambridge Econometrics, UK, 1991.

even in many OECD countries. There can also be a "carbon leakage" problem if energy taxes are not co-ordinated among countries (see page 73). In general, governments will prefer polices involving low to medium economic cost and low to medium political difficulty — assuming they are able to assess the cost and difficulty of implementing each measure.

It is difficult to be specific about the political acceptability of particular measures because of the considerable differences among regions and countries. Higher energy taxes have proved to be acceptable in some European countries, but have been strongly resisted in the United States, Canada and Australia. Generally speaking, it is easier to motivate institutional bodies than individual end-users. Unlike individuals, institutions do not vote, although they do have other ways of expressing their views. For this reason, fuel efficiency standards for vehicles have proved to be more politically acceptable than taxes on transport fuels in North America.

Types of policies and measures that might be considered

Section III of the IEA Statement on the Energy Dimension of Climate Change describes the measures available to governments to reduce greenhouse gas emissions. Technological progress can help break the links connecting economic and population growth to increased energy consumption. Energy technology which is being commercially deployed today has been partly shaped by government activities and funding over many years. In the long term (to 2050), R&D policies to speed up the otherwise lengthy process of technological development and deployment are expected to play a major role in climate change mitigation, provided governments give them high priority. In the short to medium term (to 2010), however, new R&D activities are not likely to reduce energy demand and emissions significantly because of the very long lead times involved in researching, developing, demonstrating and commercialising new energy technology. However, there is evidence that some related policies have accelerated certain technology developments and their acceptance by consumers.[3]

The cost-effectiveness of specific emissions abatement policies and measures differs significantly among regions and countries: the least-cost measure or mix of measures in one region or country will not necessarily be cheap in another. The prime determinant of cost-effectiveness is the existing infrastructure for producing and delivering energy services, such as transportation and various electricity end-uses. There are substantial differences among Annex I and non-Annex I countries[4]: the structure and pattern of energy demand, the fuel mix, the technology mix, the age and replacement rate of capital stock, import and export balances. The potential for cost-effective emissions reductions by energy users,

---

3   For example, the Golden Carrot and Energy Star policies in the United States.
4   Annex I countries are those of the OECD, Former Soviet Union (FSU) and East and Central Europe (ECE); non-Annex I countries are those of the rest of the world (ROW).

the scope for low-cost policy actions to induce those reductions and the most appropriate policy options are also diverse. This suggests that internationally agreed measures such as emissions trading or joint implementation may achieve a given reduction in emissions at world level at lower cost than through individual country action.

There is little empirical and analytical understanding of the ranking of policies and measures according to practicability and cost-effectiveness — even at the country level. Most of the economic literature has focused on the costs and benefits of specific emission-reducing actions rather than on the political difficulties associated with policies to induce individuals, firms and institutions to take those actions. Experience, however, may allow us to assert the following:

- Removal of distortions (e.g. energy subsidies) and market reforms may constitute the single largest area of low-cost government action. The scope varies enormously among countries. Some countries — especially in the OECD, Eastern/Central Europe and Latin America — are already well down this road. Others — especially developing countries and the FSU — are only just starting. In most cases, the groups receiving subsidies can wield political power to oppose their suppression and governments may have objectives other than economic efficiency in providing energy subsidies. In either case, political difficulties arise.

- There may be significant scope, particularly in OECD and transition countries, for expanded low-cost, government-supported sectoral approaches, such as voluntary agreements, promotion of audits, technical support, information and awareness programmes and, in some instances, low-impact energy performance standards.

- Economic measures, especially carbon taxation and tradeable permits, could play an important role in some countries. Potentially, they provide the least-cost mechanism for achieving significant reductions in CO_2 emissions — particularly if co-ordinated internationally — by allowing consumers to seek out the lowest-cost ways of reducing their emissions. Price signals enable individual energy consumers to make rational choices based on information that is not generally available to officials who must set regulations and ensure that they are implemented. Sensitivity to carbon value varies greatly according to the energy-related service and the sector. Responses to oil shocks show that in the power sector or in industrial stationary fossil fuel end-uses, i.e. those areas in which taxation is by far the lowest, carbon values of a few tens of US dollars per tonne of carbon would shift the merit order of existing plants burning different fuels. They would also alter the ranking of new plants in terms of per-unit present discounted value of all plant costs (capital, operating and maintenance, fuel, etc.). Gas would be favoured over coal and oil. Oil would be favoured over coal. Other factors being equal, nuclear power, hydropower and other renewable sources of energy would be favoured over fossil fuels. Clearly, however, factors other than unit cost will be involved in decisions relating to the building of new nuclear or large hydropower plants. These factors include safety, public acceptance, environmental impact and siting permission. Residential and commercial uses of fossil fuels would be affected to a much lesser extent by pricing. Unless very high carbon values were introduced, demand for transportation use and electricity generation would be even less affected.

The widespread use of economic or regulatory measures capable of inducing significant reductions in CO_2 emissions would require international co-ordination to prevent "carbon leakage." Carbon leakage occurs when a reduction in emissions in one country is offset by an increase in another country through migration of industry. This effect can also occur if fuel costs, and hence the prices of goods and services, are raised as a result of environmental policies in some countries and consumers switch to lower-priced goods and services produced elsewhere. Carbon leakage does not arise in the power generation sector except where there is substantial international trade in power.

New CO_2 abatement policies must be phased in carefully to minimise the costs of compliance and implementation. Policies that require the replacement of existing plant and equipment before the end of their expected economic lives can impose substantial economic costs. A carbon tax or system of tradeable emission certificates announced in advance and introduced over several years would allow investment in energy-supply facilities and energy-using equipment in a way that minimises costs. This approach would also send a long-term signal to everyone that such measures were permanent.

POLICIES AND MEASURES BY SECTOR

Homogenous and comprehensive economic policy measures (such as a carbon tax or tradeable permits) constitute the most cost-effective approach to reducing emissions because they minimise the cost of marginal emissions abatement across all sectors. However, in many countries significant rigidities exist in energy-supply systems and in the demand for energy-related services (see Chapter 1). These rigidities require the use of sector-specific policies that address identifiable market failures or barriers. Such actions can complement economy-wide approaches.

The following section discusses the types of policies and measures that could achieve the reductions in energy demand and emissions in the Energy Savings Case. These policies and measures could be applied economy-wide or solely at sectoral level. They could be used in a single country, a group of countries or used in all Annex I countries. Each energy-use sector (electricity, transport and stationary non-electrical) is examined, taking account of political acceptability and cost factors. The discussion focuses primarily on OECD countries.

Governments are the proper bodies to determine which policies are appropriate to the circumstances of particular countries.

Electricity supply: Measures to promote more efficient and less carbon-intensive generation

Table 4.3 sets out an illustrative range of key supply-side measures which, if implemented, could bring about significant reductions in power sector emissions.

Table 4.3
Policies and Measures Available to Governments to Reduce CO_2 Emissions in Power Generation*

Policy/measure	Impact
Introduction of or increased competition in generation**	- Reduces fuel costs - Incites search for least-cost generation options - Prevents subsidisation of fuels and plants - Opens opportunities for more energy-efficient, decentralised co-generation
Carbon taxes/tradeable permits	- Raise overall cost of fossil fuel inputs and encourage earlier retirement of high-cost plant - Encourage fuel switching in existing and new plant through price signals - Ensure flexibility in how reductions are achieved, thereby minimising marginal cost
Development of gas infrastructure	- Encourages fuel switching from coal and oil
Promotion of non-carbon-emitting fuels	- Encourages direct investment in nuclear power and/or renewable sources of energy
Audits and voluntary agreements***	- Can encourage generators to undertake emission-reduction measures they would not otherwise consider

* Excluding demand side.

** In some cases, increased competition in electricity generation could lead to higher emissions; lower prices can encourage higher electricity use.

*** Formal agreements between government and industry to improve energy efficiency or reduce CO_2 emissions below a specified baseline.

The introduction and promotion of competition in electricity supply could be a cost-effective policy for promoting increased generation efficiency. The scope for market reforms is large in many OECD countries and in FSU/ECE and developing countries. In a competitive environment, the generating industry has a strong incentive to minimise fuel costs and other operating costs through efficiency improvements. In the United Kingdom and the United States, competitive pressures have led to improvements in the operation of existing coal-fired plants and to the mothballing of older, less efficient boilers. The open and competitive Nordic electricity market has resulted in more efficient use of production capacity. Competition also encourages the search for new technology to gain a cost advantage. The availability of ample gas supplies may favour gas-fi[r]ed combined cycle technology for new full-time and part-time capacity, resulting in both higher efficiency and a lower level of carbon intensity. Competition may also promote more energy-efficient decentralised co-generation (combined heat and power) in industry and in the residential and commercial sectors. In some cases, however, where existing policies favour low or zero carbon fuels, liberalisation of the electricity sector could lead to an increase in emissions. For example, in the United States, a study by the Federal Energy Regulatory Commission (FERC) indicated that increased national competition in electricity could, under certain circumstances, result in a wider use of base-load coal-fired plants at the expense of gas-fired plants.[5] In addition, increased competition lowers energy prices, encouraging higher energy use. Consequent cost-cutting often targets R&D.

Obviously, in increasingly competitive electricity supply markets, power generators base their investment and management decisions primarily on expected financial returns. One means of ensuring that CO_2 emissions are taken into account is to introduce a price signal into generators' decision-making and operations. Taxation of fuel inputs according to their carbon content or the use of tradeable emissions permits (linked to emission limits) provide the most economically efficient ways of achieving this. At present, carbon taxes are applied to inputs to power generation in only two countries: the Netherlands, where the tax is based both on energy and on carbon content, and Norway, where over 99% of electricity is produced from hydropower. Tradeable permits for SO_2 emissions have been effective in the United States in bringing about reductions in a least-cost fashion.[6] CO_2 emissions limits linked to a tradeable permit scheme could be applied in or among some OECD countries[7] or, more effectively, at a broader international level. The problem of "carbon leakage" would need to be addressed if carbon taxes or tradeable permits were not deployed in an internationally co-ordinated manner (see page 73).

Natural-gas-fired combined cycle technology is in many cases the most economic generation option. This could be a significant factor in developing countries, where the gas sector is relatively underdeveloped. In most OECD and FSU/ECE countries, the gas infrastructure is already very extensive. In the United States, the introduction of competition in the gas industry was instrumental in eliminating bottlenecks in gas infrastructure and increasing the overall use of gas. In other countries as well, opening the gas industry to competition can be an effective way to remove bottlenecks in infrastructure (e.g. the UK-Continental Europe interconnector), bringing more supply to the market and increasing security of supply.

---

5   Environmental Impact Statement on Promoting Wholesale Competition Through Open Access Non-Discriminating Transmission Services by Public Utilities (RM95-8-000), FERC, 1995.

6   The success of the SO_2 tradeable permit system has prompted northeastern American states to develop a similar programme for NO_x emissions, to be introduced in 1999.

7   Plant-specific or sectoral emissions ceilings for a range of air-borne pollutant emissions are already widely used in OECD countries.

Further investment in nuclear plants for commissioning before 2010 may be possible in Japan and in some other OECD countries, provided the problems of political acceptability and waste and safety concerns can be adequately addressed. There is potential for increased nuclear capacity in FSU/ECE, but it is questionable whether the rate of growth of electricity demand will justify increased investment. Nuclear power could also contribute significantly to baseline emissions reductions in some developing countries, most obviously in China, India and other fast-growing Asian countries.

Governments in OECD countries can probably do little more to encourage the development of hydropower resources. Such development is, for the most part, constrained by cost or environmental considerations and by the availability of suitable sites. However, large hydro resources exist in non-OECD countries.

The share of non-hydro renewable resources in generation is increasing. However, their costs are unlikely to fall sufficiently for them to compete on a large scale with established fuels in the near to medium term without a high carbon tax or high tradeable emission certificate price.

Additional government support for development and deployment of energy technology would probably not result in any significant overall reduction in CO_2 emissions until after 2010.

Voluntary agreements based on objective auditing procedures (beyond those already negotiated), could contribute to reducing CO_2 emissions in OECD countries — especially if they are backed by a credible threat of regulatory or fiscal measures[8] and if the problem of carbon leakage can be successfully addressed. However, their potential is probably limited and they may distort normal business decision processes.

Electricity demand: Measures to curb demand growth

Governments have relied mainly on information campaigns to curb the growth in electricity demand since the first oil shock. Canada and the United States have the most comprehensive programmes for minimum energy efficiency standards, dating from the late 1980s (although actions at state and provincial level were taken earlier). Over twenty different appliances and equipment items are subject to standards harmonised between the two countries. In Japan, the Law Concerning Rational Use of Energy was enacted in 1979 and partially revised in 1993. This law provides for energy efficiency improvements in:

- ■ factories (principally "designated" factories), through energy audits, improved energy management, energy consumption reporting and rationalisation plans;

---

8 As indicated in the recent in-depth Review of the Energy Policies of the Netherlands (IEA/OECD 1996), there is a trade-off between the tightness of the target for reducing CO_2 emissions and its economic acceptability.

- buildings, through regulations, guidance and advice;

- equipment, through recommendations to manufacturers and importers of energy-using equipment concerning energy efficiency standards; (New guidelines for air conditioners and passenger cars were announced in 1993; for fluorescent lamps, TV receivers, copying machines, electronic computers and magnetic disc drives in 1994; new guidelines are expected for refrigerators in 1997.)

The first EU directive for minimum energy efficiency standards for appliances, under the SAVE programme, adopted in 1991, covered liquid and gas-fired boilers in the 4-400 kW range. All EU Member states are required to enforce this directive by 1997. More recently, the EU adopted a minimum efficiency level for refrigerators. Denmark, Norway, Switzerland and the Netherlands have more extensive appliance standard programmes than the EU requirements.

In all countries, electricity consumption has risen as demand has risen for the services that electrical equipment can provide. Rising income has been a major cause. New electrical products have also appeared on the market. Although increases in electrical appliance efficiency reduce the rate of growth of electricity consumption, they also lower the operating costs of appliances and release some real income that can be spent on additional electrical services. This offsetting effect is termed the "rebound effect." The short to medium term price elasticities in demand for electricity are generally low and the rebound effect for electrical appliances is small. It is smallest for those appliances for which extra energy efficiency can only be achieved at higher capital (or running) cost. The impact of stricter energy-efficiency standards on average electricity use per appliance will depend on their replacement rate. The full effect would take 10 to 20 years to be felt.

Although there is evidence that electricity demand in a number of markets is sensitive to price (e.g. in electricity-intensive industries, such as steel and aluminium, and in contestable fuel markets, such as space heating), total electricity demand is not very sensitive to price changes in the short term. Consequently, although the real price for electricity in OECD countries rose by about 17% during the period from 1978 to 1985 before returning to its former level, there is little evidence that this increase resulted in behavioural changes in OECD regions as a whole. And, while carbon taxes may strongly affect the mix of fuels in power generation, such taxes would have a smaller proportional effect on the final price of electricity because of the weight of non-fuel costs in final prices (e.g. capital, operating and maintenance costs of generation, transmission and distribution).

Rigidities in the electricity sector arise because demand for electrical appliances rises sharply with income, while demand responds weakly to changes in electricity prices. The slow process of capital stock turnover also contributes to rigidity. Policies that increase the energy efficiency of new equipment by promoting the use of new technologies are likely to be more effective than policies that raise the prices of fossil fuels used in electricity generation. But even those policies will be slow to show effects because of the slow process of changing capital stock. Better promotion and marketing of energy-efficient appliances and equipment could help a lot. Anecdotal evidence suggests that opportunities are being missed, for example, to promote energy-efficient refrigerators and heat pumps in some countries.

Other measures that could have additional, although modest, impact on electricity demand include:

- **Information dissemination, education and awareness programmes:** Most OECD countries and some transitional and developing countries already have such programmes. There is scope for introducing and expanding such activities at zero or negative net economic cost. Experience with industrial audits in the United States over the past two decades suggests that such programmes can bring new technology and practices to the attention of management and that about half of the cost-effective actions for reducing energy consumption identified in the audits are implemented.[9]

- **Voluntary actions and programmes in the industrial and service sectors:** Such programmes have been implemented or are in development in most OECD countries. They are sometimes deliberately combined with other policy measures, such as financial incentives and energy audits. It is uncertain to what extent voluntary agreements encourage energy-efficiency improvements compared with what would have happened in their absence. Their apparent effectiveness may vary according to how closely they are monitored.

- **Public procurement rules** that encourage the accelerated development and production of energy-efficient products.

Demand Side Management (DSM) strategies to promote energy efficiency in electricity end-use will need to be adjusted in the face of structural and regulatory reforms. Firms in competitive electricity markets may find DSM a profitable service to offer as a way of attracting and retaining customers, but competition is likely to reduce the role of DSM in curbing demand growth in the medium term. There will, however, always be a role for independent energy service companies that can make a profit by sharing the value of the energy saved with their customers.

Transport: Can the policy impasse be broken?

OECD governments have found it very difficult to find cost-effective and politically acceptable ways to reduce or limit the growth of emissions from transport. The problem is compounded by the complexity of the factors which stimulate demand for transport services (including incomes, demographics, land-use planning and lifestyles) and by the complexity of the technical and

---

9 IEA/US DOE, *Industrial Energy Efficiency: Policies and Programmes, Conference Proceedings - Session 1A* (1994).

operational interactions among the determinants of energy consumption for transport.[10] It is nonetheless apparent that long-term policies that have raised the cost of owning or using a vehicle or that have mandated fuel-efficiency improvements for new vehicles have had some success in tempering the rise in transport-sector energy use.

To exploit the cost-effective opportunities in the transport sector, a mix of two key measures is probably necessary: average fleet fuel economy standards and taxation of fuel, road use and/or vehicles and aircraft. In 1997, three IEA Members (Sweden, the Netherlands and Germany) introduced environmentally differentiated taxes on vehicle registrations, and Denmark added a tax that rises with the fuel consumption of cars (above a certain threshhold). The Netherlands raised diesel excise taxes and fuel taxes in general, but lowered fixed taxes slightly. In some cases, improved average vehicle fuel efficiency may be achieved through voluntary actions by car manufacturers to avoid mandatory standards. Mandatory or voluntary standards and taxes affect emissions by improving the fuel efficiency of the national vehicle fleet through their impact on cost. Improvements in overall fuel economy can result from improvements in vehicle efficiency and a shift in consumer choice towards smaller, less fuel-intensive cars.

As in the case of electricity use, Chapter 1 shows that demand for petroleum fuel for mobility is strongly dependent on economic growth. In addition, motorists are turning to diesel-fuelled cars in many countries. They are attracted by diesel's lower running costs per km, brought about by lower pump prices for diesel fuel compared with gasoline and the higher on-road efficiency of diesel cars. (Other differences between gasoline and diesel-fuelled cars, such as car cost, noise, weight, etc. are now small.) In 1995 in France, the average annual distance travelled by a diesel car was 21,500 km and by a gasoline car, 11,700 km. However, an analysis of motorists who switched from gasoline to diesel cars showed that, on average, they already drove 15,500 km per annum before switching and they drove 20,000 km after switching. This suggests two things: first, long-distance motorists are attracted to the lower running costs of diesel cars and, second, the greater efficiency of diesel cars lowers motoring costs and encourages greater use.

Estimates of transport fuel price elasticities of demand for OECD regions, incorporated in the IEA's World Energy Model, imply that a large increase in fuel price taxation would be necessary to reduce fuel demand and emissions significantly. Experience with mandatory average fuel efficiency standards is limited to North America, where Corporate Average Fuel Efficiency (CAFE) standards have contributed to improvements in fuel economy (see Chapter 1). Although the CAFE programme has been criticised for its inflexibility and its redistributional effects (car manufacturers are not affected equally), it has proved to be more politically acceptable than increased fuel or vehicle sales taxes.

---

10 These issues are discussed in *Transport, Energy and Climate Change*, IEA/OECD, to be published.

The appropriate level of taxation and type of standards would vary considerably among countries:

- In OECD countries of Europe and in Japan, fuel and vehicle tax levels are already high (with a notable exception for so-called "company cars"), making further increases politically unattractive and less cost-effective than in other regions. Fuel efficiency standards, possibly involving "feebates,"[11] could be a major element of a politically acceptable, least-cost reduction strategy for transport sector emissions, supplemented by tax increases over time in some countries.

- In North America, Australia and New Zealand, mass transit is not as well developed as in Europe and the scope for higher taxation affecting passenger cars has been constrained by lobby groups (including consumers, oil companies and manufacturers) who resist any direct measures that may raise the cost of driving or damage international competitiveness. In this context, the introduction or tightening of standards for light and heavy road vehicles (such as U.S. CAFE- type standards) may have more chance of being implemented.

- In the FSU/ECE and developing countries, fuel and vehicle taxation, which could be justified largely on revenue-raising grounds, could be the principal way of exploiting cost-effective potential.

It is likely that policies for the transport sector as a whole (rather than specifically for energy used in transport) will most affect CO_2 emissions in the transport sector. The pricing of road use to limit congestion in cities, or to derive revenue from the use of motorways, would incidentally help reduce energy use and emissions, provided traffic were reduced and not simply diverted. Experience in Norwegian cities suggests that even small charges can reduce road traffic. Clearly, higher charges could be more effective, but they would be unpopular. The degree of the unpopularity of congestion charges would depend on improvements in public transport introduced at the same time. Road pricing is currently used in Singapore and is under consideration in a number of IEA countries besides Norway.

Alternative transport fuels are unlikely to play a significant role up to 2010, because of their cost and lack of infrastructure. Biofuels, liquefied petroleum gas (LPG) and compressed natural gas could, nonetheless, contribute modestly to emissions reductions in a few countries. In the longer term, alternative fuels offer prospects for lowering greenhouse gas emissions from the transport sector, although their costs remain high.

---

11 In a feebate system, purchasers of the most efficient vehicles receive a tax rebate, while purchasers of less efficient vehicles pay a tax. The zero point of the scale would be set at the average new vehicle fuel economy, adjusted periodically for changes in fuel economy, so that the overall tax burden is zero. It should be recognized that such a system would bring the same average result as an average fleet standard. In the United States, the low price of imported Japanese cars played a role similar to a feebate.

Stationary fossil fuel uses: A range of complementary measures needed to address a diversity of uses

As demonstrated in Chapter 1, energy price signals, changing industrial structures and international migration of energy-intensive industry have contributed to stabilising the use of fossil fuels, particularly oil, in the industry and service sectors in OECD countries. Policies for further reducing energy use in these sectors could involve a range of complementary measures (see Table 4.4). Market reforms that result in full-cost pricing and removal of subsidies could stimulate more efficient energy use where significant distortions exist, as in the FSU/ECE. Taxation could in many countries be a cornerstone for policies to reduce stationary fossil fuel use. This cornerstone could be reinforced with a range of sectoral measures, including voluntary agreements, government-funded technical support and information programmes, building codes and standards. Close co-operation between industry and government could help bridge information gaps, especially if linked to government-sponsored technical advice, information and energy audit programmes. Such co-operation could also accelerate the responsiveness of industry to technological change.

All IEA countries have building regulations or codes that specify minimum standards of construction and/or insulation to secure desired levels of comfort and efficiency of energy use. Most of these regulations have been in place since the early 1970s. Many of the standards were tightened in the 1990s. Given that the design features, choice of materials and methods of construction required for high energy efficiency in buildings are much cheaper than modifications after construction, tighter building regulations represent a very effective method of improving energy efficiency.

The industrial and service sectors are heterogenous in nature and involve a wide range of energy-using equipment. The imposition of minimum energy efficiency standards thus poses an enormous technical challenge. Nevertheless, such standards are applied in a number of IEA countries. For example, the U.S. Environmental Protection Act contains mandatory efficiency standards for electric motors and other categories of industrial equipment, such as fluorescent and high-intensity lamps and large heating and cooling systems. In Japan, too, a number of equipment items are subject to energy-efficiency standards.

In the FSU/CEE and in some developing countries, the re-equipping of factories and refurbishment of buildings are likely to make a major contribution to the reduction of energy intensities and pollutant emissions. These processes can lead to even greater energy savings if adequate advice is available and followed and if building and other regulations are enforced. A business environment that favours foreign direct investment in these countries will bring with it efficient equipment, modern construction techniques and management methods that will lead to further energy saving. This is another example of how wider sectoral policies, in this case industrial and commercial restructuring and modernisation, can contribute to saving energy.

Table 4.4
Policies and Measures Available to Governments for Curbing CO_2 Emissions Growth From Stationary Uses

Policy/measure	Impact
Energy market reforms	- Raise end-user prices through removal of subsidies and promote more efficient energy use
Carbon taxes/tradeable permits	- Raise cost of fossil fuel inputs
	- Encourage fuel-switching in existing and new plant through price signals
	- Ensure flexibility in how reductions are achieved, thereby minimising marginal cost
Audits Voluntary agreements	- Encourage end-users and product manufacturers to undertake measures that reduce emissions that they would not otherwise have considered
Technical assistance/information	- Identify and encourage changes in practices and investment that reduce emissions
Building codes	- Guide or mandate the way in which buildings are built or renovated to limit energy use for heating and cooling
Appliance/equipment standards	- Encourage or mandate improved energy performance in lighting, heating, cooking appliances and electric motors through product innovation, design and retooling

In industries or companies where energy is a major input, the scope for any measures that significantly raise the cost of energy is limited by the problem of carbon leakage, *unless the level of the measures is small compared to the cost of transporting competitive goods from foreign suppliers not subject to the measures, or unless the measures are introduced internationally.* Indeed, most of the current carbon tax schemes include exemptions for energy-intensive industries to protect them from foreign competition. Unless policy is sufficiently co-ordinated among countries competing in international trade, it must carefully differentiate between those subsectors of industry which are the most energy or carbon-intensive and which are exposed to international competition (steel, aluminium, fertilizers and petrochemicals) and the rest, like the service sector, where energy is not a key input. But this leads to the obvious paradox of exempting the most energy-intensive processes from measures to curb energy consumption.

SUMMARY

It is important to distinguish between three main types of energy efficiency policies:

- those that provide information on how to save energy;

- those that limit consumer choice or require changes in energy consumer behaviour;

- those that involve changes in energy prices.

It is also important to realise that other policies, e.g. industrial restructuring or non-climate change environmental policies, or transport policies, such as congestion charges that reduce traffic rather than divert it, also bring about energy saving.

The potential for realisable no-regret, cost effective energy saving is very uncertain. Current IPCC estimates suggest gains of 10%-30% on baseline trends over the next two to three decades. This potential will be realised only if policies and measures can be found and implemented to overcome existing barriers. Political difficulties will limit the size of the achievable no-regret potential.

Policies to achieve significant energy savings are likely to involve both low to medium economic cost and low to medium political pain.

IEA Member countries have deployed a wide range of energy efficiency policies over the last twenty years. Some have introduced measures to raise the fuel efficiencies of new cars (the United States, Canada, Japan and Germany). Many countries introduced tighter building regulations and stricter energy efficiency standards for appliances during the 1990s. Nonetheless, the policies introduced so far have resulted in much less energy saving in the IEA regions than would be required to reduce CO_2 emissions in 2010 below their 1990 levels.

Governments in some countries may find that even low-cost policies may be politically difficult to implement, e.g. removal of subsidies for energy production or exposing domestic producers to competition from imports. Electorates need to be persuaded on the basis of broad social benefits, such as reduced urban pollution and traffic congestion. They will also need to be satisfied on issues of equity.

Low-cost policies include regulatory reforms (particularly in the electricity and natural gas sectors) and the removal of market distortions (e.g. energy subsidies). Other policies and measures, such as voluntary agreements, relatively undemanding energy efficiency performance standards, energy efficiency information dissemination, technical assistance and education/training programmes, are also generally low-cost. More ambitious targets for reducing emissions will involve higher costs and may require carbon taxes, tradeable permits and more stringent standards and regulations governing fuel use and emissions.

Emissions reductions are likely to be more difficult to achieve in the transport and electricity end-use sectors than in stationary end-uses or electricity generation. In the first two sectors, the fossil fuel cost element is only a small part of the final fuel price. The price elasticities of demand for these fuels are low. In these circumstances, very large taxes on the basic fuel cost would be required to have any significant impact on fuel use. Alternatives to the use of high taxes on the basic fuel cost include:

- Transport policy reform packages, including congestion charges in cities, motorway tolls, road taxes differentiated by size of vehicle, announced increases in the cost of road travel and provision of alternative public transport. (These are policies for transport rather than for energy use, but they will also save energy and reduce CO_2 emissions.)

- In the electricity end-use sector, policies that encourage the design and production of more energy-efficient appliances, coupled with minimum energy efficiency standards, could help offset low energy price responsiveness.

Economic measures potentially provide the least-cost mechanism for achieving significant reductions in CO_2 emissions, particularly if co-ordinated internationally. An internationally agreed economic instrument, such as an emissions trading system, could complement measures implemented by individual countries.

5

LONGER-TERM FACTORS AFFECTING CLIMATE CHANGE POLICIES

INTRODUCTION

Beyond 2010 lie a host of uncertainties that cannot be resolved in 1997. The range of such uncertainties is best dealt with by employing different scenarios.

A number of longer-term issues that affect the near-term development of climate change policies can be identified. These issues are discussed below.

TECHNOLOGICAL DEVELOPMENT

Among the important long-term changes affecting most aspects of economic life will be the pace and direction of technological development. For the energy-related aspects of climate change the vital issues are:

■ reducing the energy use of equipment for given levels of service;

■ improving the energy efficiency of energy transformation processes;

■ identifying energy transformation processes that consume a wide range of inputs (e.g. waste matter) and at the same time have lower pollutant emissions (e.g. gasification);

■ developing practical means for capturing, storing and disposing of pollutants from industry and from energy production and transformation processes;

■ widening the range of low-emission and no-emission forms of energy;

■ lowering the unit costs of all such technologies.

The adoption of some of these technologies may require government support to overcome the barriers discussed earlier. The faster the rate of development of relevant technologies, the greater will be the scope for achieving climate change objectives.

The direction and pace of this development can be affected by economic factors. The pressure on private companies to survive and increase market share in a competitive environment impels technical development, both to reduce unit costs and to extend the range of products. Salient examples exist in the pharmaceutical and computer industries. Such pressures would be enhanced by strong regulations limiting the sale of equipment with low energy efficiency. A similar effect can be achieved by adding identifiable and quantifiable environmental costs to the prices of fuels and, hence, of final products. Examples of ways of including environmental costs in prices are carbon taxes and tradeable emissions permits. Clearly, the removal of subsidies for the production and use of fuels emitting greenhouse gases would be a major positive step. The transfer of new technologies — as well as advanced training, management techniques and manufacturing methods — to developing countries, where the growth of CO_2 emissions is expected to be fastest, would be encouraged by Activities Implemented Jointly by Annex I countries and developing countries.

Government funding of selected Research, Development and Demonstration (RD and D) projects can also assist by directing activities towards government goals. Current trends show declining RD and D budgets, in both the government and private sectors of IEA Member countries, and an increasing focus on near-term aims. At present, it is unclear that sufficient innovative climate-friendly technologies will be available to meet the demands of longer-term climate change commitments. The IEA seeks to reduce the costs of RD & D by encouraging international co-operation in the funding and execution of research projects. The Agency seeks to enhance technology transfer through increasing the participation of key developing countries in such research.

FACTORS AFFECTING ENERGY DEMAND

Significant proportions of the capital stocks of vehicles, plant, machinery and buildings existing in 2010 will be built over the next 13 years. Their type, construction and operation will be influenced by policies and measures adopted in Kyoto at the end of 1997 and subsequently. Passenger cars are a prime example. (The average life of a car is 10 to15 years.) Opportunities for achievement are lower for assets with longer lives, and they vary widely by type of asset and country of location. For many types of plant and machinery with average lives of 10 to 25 years, 50 to 100% will be built between 1997 and 2010. For buildings which have an average age of 80 years, the proportion constructed after 1997 will be in the region of 16%. In addition, refurbishments of older stock will be induced by economic and/or political necessity. Thus, governments have the opportunity now to affect a sizeable proportion of the capital stock that will exist in 2010 and beyond.

The range of climate change policies and measures likely to be introduced over the period to 2010 will vary from country to country to suit individual circumstances. Market-based instruments (e.g. tradeable permits or carbon taxes) could play an important role, in conjunction with regulations and other policies, where markets do not exist, where competition is weak or where sensitivity to a carbon value would be low. Reducing CO_2 emissions across countries at the lowest marginal cost implies some form of joint implementation among rich and poor countries, perhaps ultimately a system of tradeable permits, if the necessary terms and arrangements can be successfully negotiated.

One consequence of the use of market instruments is that the end-user price of each energy form will progressively reflect the external environmental costs associated with the production and use of that energy source. Regulations designed to reduce CO_2 emissions should incorporate compliance costs that approach these external costs.

For countries with high and rising per capita incomes (i.e. many of the OECD and fast-growing Asian economies), the possible onset of saturation in energy demand, or a slowing of demand growth, is an important issue. But these effects can be difficult to identify because of the many other factors affecting energy demand in each sector. The timing of these saturation effects is uncertain, but they could become significant after 2010. They need to be taken into account by governments when evaluating climate change policies for the longer term.

- **In transportation**, growth in total annual distance travelled by road per person, private vehicle ownership and vehicle use could begin to slow by 2010. Contributing factors will be the costs and discomfort of travel, such as congestion in urban areas, and new work habits as telecommunications and computers substitute for personal transportation. Some levelling out in car ownership and distance travelled per car is already apparent in some countries.

- **In electricity use**, as the share of households owning specific electrical appliances rises towards 80 to 90% in some countries, the rate of growth of electricity demand for use in those appliances will slow. It will approach the rate of growth in the number of households, offset by improvements in the electricity efficiency of those appliances. Electricity demand increases through the ownership of multiple appliances by the average household, but this demand is usually offset by increasingly lower utilisation of each appliance. In some cases new appliances phase out old ones, in the way audio tapes largely replaced discs and compact discs largely replaced tapes. A careful analysis by type of electrical appliance is required to distinguish evidence of saturation in electricity demand. Air conditioning shows little sign of saturation.

  Clearly, the increasing range of electrical appliances is a major factor driving growth in electricity demand. The substitution of electricity for other fuels is also a contributing factor. Further shifts to electricity and away from other fuels could be seen after 2010 — e.g. electric vehicles or electric heating in energy-efficient buildings with sophisticated controls.

- **In heat** provided by fossil fuels, some saturation effects have already occurred. Temperatures in buildings have peaked for most households, as well as public buildings. The number of homes with central heating is approaching 100% in a number of IEA countries. There will continue to be some increases in demand as low-income households approach normal comfort levels, but for the bulk of the population, energy demands for heat will level off in high-income countries. Saturation in the demand for heat used in industrial processes is less clear. Overall, fossil fuel end-uses could level off or even decline if more combined heat and power (CHP) were used.

There is evidence that the growth in consumption of non-commercial biomass fuels, such as wood, is slowing in developing countries, while the growth in consumption of commercial fuels continues unabated. Such factors need to be identified and clarified so that they can be taken into account in preparing future projections of energy demand and CO_2 emissions.

POSSIBLE DEVELOPMENTS IN ENERGY TRANSFORMATION

Liberalisation and deregulation of electricity and gas markets are likely to continue in all countries. In developed countries, the main emphasis is on competition or a high degree of regulation to compensate for lack of competition. Developing countries need to attract private investment funds (instead of government finance), mainly through privatisation. Both the enhancement of competition and policies to attract inward investment imply a trend towards adequate returns on capital investment and full-cost pricing. This, in turn, will require the phasing out of energy subsidies. Both will provide a firm basis for the introduction of market-based climate change policies. Private financial flows to developing countries have more than quadrupled from $50bn in 1991 to over $200bn in 1996. Official financing (governments, World Bank and IMF) remained static at around $50bn.

Power generation is the area offering the greatest short-term flexibility for fuel switching. This is due to part-loading of many power stations during off-peak hours. In addition, the construction of new power stations, the refurbishment of existing stations and the decommissioning of old ones provide longer-term opportunities for reducing CO_2 emissions. Following the two oil crises of 1973-74 and 1979, existing power stations were used differently. Oil-fired plants were used less and gas-fired stations were used more. New power stations were built. This new construction then slowed because of the existence of surplus generating capacity in many parts of the developed world. The generating plant reserve margin (excess of generating capacity over peak generation rate) for OECD countries as a whole has been close to 30% since the late 1980s. Figure 5.1 indicates the variations in generating plant reserve margin for a number of OECD countries from 1985 to 1994. The margins in Europe and in the United States are higher than average. Those in Japan are lower. Evidence suggests that reserve margins are also high in much of Eastern Europe and the FSU. Investment in new generating plants in these regions may be expected to be delayed until this surplus falls to zero, delaying this potential for reducing CO_2

emissions — unless new generating possibilities arise for which the total generating cost (including return on capital investment) is less than the operating and fuel costs of existing plants. In this latter case, a new plant could become economic in displacing an operating plant.

Figure 5.1
OECD Generating Plant Reserve Margins

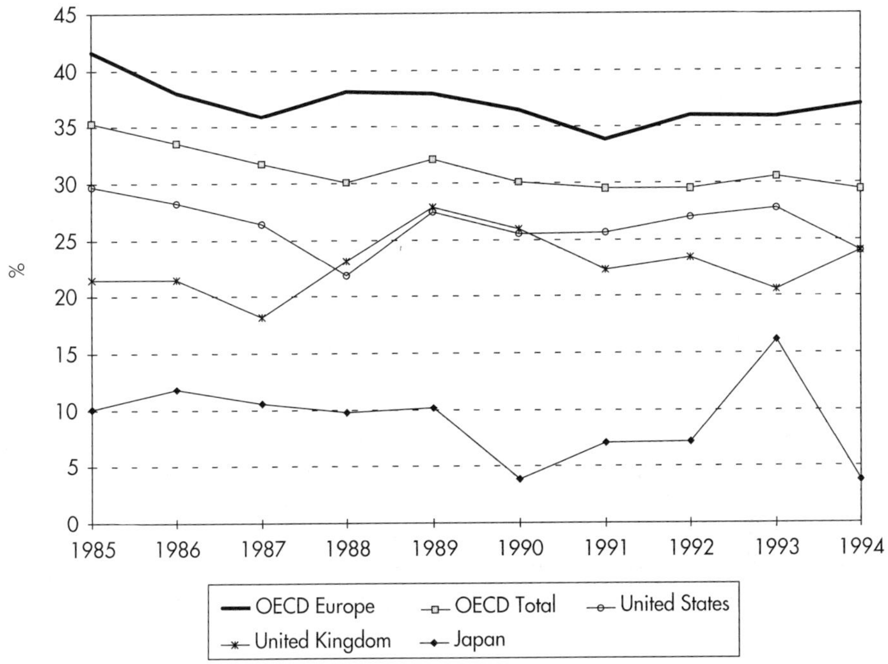

Note: Reserve margins based upon available capacity at peak, if data are available. Otherwise nominal installed capacity is used.

Cogeneration of electricity and heat for industrial enterprises, for district heating systems or for individual buildings offers higher thermal efficiency, hence lower CO_2 emissions, per unit of electricity and heat (combined). There is some concern that district heating systems may become uncompetitive in some countries when their owners lose their monopoly and face lower prices from liberalised electricity and gas utilities. The position of cogeneration in the marketplace would improve considerably under price-related policies that aim to include the external costs of CO_2 emission (carbon value) in the product price. Supplementary efforts to promote cogeneration are under consideration by the European Union (EU) Commission. Ideas include: an EU target for combined heat and power (CHP) capacity, an updated review of the market barriers to CHP, greater use of funding for non-EU countries (via the Phare and Tacis programmes), greater co-ordination of statistics on cogeneration among EU countries, and the promotion of new CHP technologies.

Coal use will be most affected by the introduction of policies to reduce CO_2 emissions because of coal's high carbon/energy ratio. Switches from coal use will occur mostly in power generation and heat production, where alternative fuel supplies are available. Integrated Gasification and Combined Cycle (IGCC) technology promises higher thermal efficiency in the production of electricity from coal, once the remaining problems of hot gas clean-up have been resolved. It is unlikely, however, that this technology will be competitive with combined cycle gas turbine (CCGT) plants at *current* fossil fuel price levels. The latter enjoy lower capital cost, shorter construction time, availability in smaller units and a lower carbon/energy ratio. In the switching process, coal will lose out to gas unless gas prices rise considerably.

Because oil is easy to store, it is the most versatile fuel. Oil use for electricity and heat production frequently occurs at the margin of operation. The carbon/energy ratio of oil lies between that of gas and coal. Oil could remain a lower-cost fuel option for countries that are remote from natural gas supplies, e.g. countries that rely on liquefied natural gas. Part-loaded oil-fired plants provide security because they can generate additional electricity if gas supplies are restricted. They are also a convenient place to reduce oil use if oil supplies are restricted. The responsiveness to changes in GDP and fuel price is much smaller in the transport sector than in the power generation and heat sectors.

Unless there is a dramatic rise in the supply cost of natural gas, it is likely that gas will make substantial inroads into coal use in the flexible areas of power generation and heat production. Oil, too, is likely to yield ground to gas in some countries, but oil use will be sustained in other countries to provide diversity, flexibility and security of supply, especially where oil is in direct competition with coal. These changes should be well underway by 2010. A key question will be how much more gas will be used beyond the baseline estimate.

Of the two current carbon-free power sources, hydropower provided 19% and nuclear power provided 18% of world electricity generation in 1994. Although they emit no CO_2 from their main equipment, some emissions arise from the full fuel cycles of each technology. In addition, each technology has other non-greenhouse gas environmental impacts.

In OECD countries, no major new hydroelectric power sites are available for development without overcoming serious social or environmental problems. In non-OECD countries, undeveloped hydro resources are substantial. Development of these resources poses many problems, including sedimentation, damage to fish stocks and population displacement. Political issues of ownership and control and, in some cases, religious attitudes about the use of water must be faced. But clearly, increased use of hydro power could make a significant contribution to CO_2 emission-free power generation in non-OECD countries.

Some other renewable energy technologies are now economic under favourable conditions, e.g. wind power, bio-fuelled CHP and solar energy. The unit costs of renewable energy technologies are falling. The question is when will they become more generally competitive with conventional fuels. Some of these technologies, such as wind, wave and tidal power, also have environmental impacts. Many IEA countries support the development of renewable energy projects through subsidies and other means. Allocating subsidies by tender can be an efficient way to choose amongst alternative renewable projects. Activities implemented jointly could be used in developing countries, where some 70% of investment in new generating plant is expected to take place.

Nuclear power has the potential to provide large quantities of electricity without emitting greenhouse gases. Apart from Japan and France, OECD countries are either opposed to the use of nuclear power or, for a variety of reasons, unlikely to permit new nuclear plants to be built for the foreseeable future. Issues include fear of accidents involving the release of radioactive material, failure to agree on locations for the disposal of radioactive wastes and concern over the magnitude of long-term financial nuclear liabilities. Advanced designs exist for smaller, safer nuclear power stations. However, because few nuclear power stations have been built over the last 10 to 15 years or are planned over the period to 2010, the pool of scientists, engineers, technicians and industrial resources which is necessary to sustain a major building programme of nuclear power stations is contracting. A substantial educational and technical training programme would be required to start such a programme if needed after 2010. The construction of nuclear power plants is being considered in some of the major developing countries and in countries of the FSU and ECE.

All energy sources and transformation technologies that can contribute to reducing greenhouse gas emissions have strengths and weaknesses. The importance of using such technologies (and the policies and measures needed to put them in place) will need to be balanced against the perceived disadvantages. It is clear that putting a value on carbon would favour the adoption of hydro, nuclear and renewable power sources in both OECD and non-OECD countries, but the opportunities are greater in non-OECD countries.

THE FUTURE FOR FOSSIL ENERGY SUPPLIES

Growth in coal demand is expected to slow up to 2010 and beyond.[1] This will be a reduction in the *rate of growth for coal demand*, not an absolute reduction. Gas demand is expected to rise, mainly as a result of changes in the relative unit costs of power generation and heat supply and the expected rate of capital stock

---

1   Developments in coal markets are discussed in *International Coal Trade: The Evolution of a Global Market*, IEA/OECD, to be published.

turnover in these two areas. There will be continued pressure for additional policies for CO_2 emission reduction that take account of the carbon content of fuels. Such policies could bring nearer the time when demand for coal begins to decline, especially if developed and developing countries act together. Much will depend on the arithmetic of rates of fuel substitution relative to rates of production of oil and gas as recoverable reserves are utilised.

Subsidised coal production is expected to cease in Japan soon after 2000. Although questions remain about the rate of progress in Germany and Spain, coal production subsidies in OECD countries are expected to be phased out by 2010. Information about levels of financial assistance to coal production in non-OECD countries is not available. Many coal mines in China, the FSU and Eastern Europe are believed to be unprofitable. Coal imports into India are currently restricted to protect the domestic industry, but removal of the restriction is under discussion. The removal of subsidies for the production of coal and other fossil fuels will be an important step toward reducing CO_2 emissions. The difficulties faced by governments in reducing these subsidies concern employment and regional regeneration. These difficulties are not easily or speedily overcome.

Analysts' views on ultimate recoverable reserves of conventional oil vary widely. Laharrère, Campbell and Ivanhoe are conservative, putting these reserves at some 1800bn bbls, compared with the United States Geological Survey (USGS) figure of 2300bn bbls and the range used by the U.S. Energy Information Administration, Department of Energy (EIA/DOE) of 2100 to 2800 bn bbls. Adelman and Lynch are more optimistic. Assuming a 2% growth rate for oil demand and taking the common assumption that oil production begins to decline once half of reserves have been produced, these reserve estimates suggest dates of peak production ranging from 2000 to 2020. All of these estimates are highly uncertain. In particular, estimates of the timing of peak production could be extended if unforeseen technological advances and high oil prices increase the amount of recoverable oil reserves. Alternative liquid fuels are available from unconventional sources in large quantities, but possibly at higher costs. (Some tar sands are already being developed today in Canada.)

Masters (U.S. Geological Survey, 1994) gives a range of 9790 to 16909 10^{12} cu ft. for the ultimate recoverable reserves of natural gas. These estimates are based on cumulative production to 1/01/93 and the identified remaining reserves and estimates of undiscovered gas. Masters' lower figure includes a 95% confidence estimate for undiscovered gas and his higher figure, a 5% confidence estimate. Assuming the continuation of 2.9% growth per annum in world gas demand foreseen in the IEA 1996 *World Energy Outlook*, half of these total gas reserves would be consumed between 2020 and 2035. Just as for oil, all of these estimates for natural gas are highly uncertain. Exploration has been less intense for gas than for oil. Further large finds of gas, developments in technology and higher prices could delay peak natural gas production beyond the estimated dates cited above. If major climate change policies brought about a rapid increase in gas demand that was not anticipated by gas producers, bottlenecks could occur in the gas supply infrastructure, bringing sharp increases in the gas price until the bottlenecks could be cleared.

The IEA (following ideas originally suggested by Michael Grubb of the Royal Institute of International Affairs in London) has calculated the amount of fossil fuels that would need to be burned to increase the CO_2 concentration in the atmosphere from the 1994 IPCC figure of 358 ppm to a limit of 550 ppm or twice the pre-industrial level. (This target was included in Proposal 1 from the EU, put to the Sixth Session of the Ad Hoc Group on the Berlin Mandate, 3-7 March 1997, in Bonn.) The calculations suggest that if all the recoverable oil and gas reserves were to be burned (using mid-range reserve estimates), coal consumption would need to be rapidly curtailed in the next century to meet such a target.

CLIMATE CHANGE POLICIES AND ECONOMIC GROWTH

World economic growth and the relative growth rates of different regions will be major determinants of future energy demand. Future economic growth rates for different world regions are interrelated and hard to project. The major determinants are:

- population growth and urbanisation;

- growth and productivity of the labour force;

- growth and productivity of capital stock;

- land use and productivity of land;

- further discovery and development of natural resources;

- technological developments;

- residual productivity growth (not already included above);

- financial transfers;

- developing trade.

The impacts of these factors on the economic growth rates of the main world regions are analysed by the OECD in "Towards a New Global Age: Challenges and Opportunities." [2]

---

2   "Towards a New Global Age: Challenges and Opportunities," OECD, 1997.

Energy availability (and security of supply) is important for economic growth. The expansion of infrastructure, including the electricity grid, and the availability of fossil fuels through the provision of roads, railways, waterways or pipelines are important determining factors for economic growth in a number of developing countries. The rate of economic growth in these developing countries will depend on their removing constraints on infrastructure development.

In the context of growing demand for energy, the major question is how governments will address the climate change problem. Three extreme responses may be envisaged:

A. No further policy developments. Industrialised nations do not launch any significant policies to reduce greenhouse gas emissions. This could be because the Kyoto negotiations fail to reach agreement or because some parliaments refused to ratify a protocol agreed in Kyoto, feeling that the burden of the policies is not equally shared, or because of concerns about impacts on the world economy. Failure to launch new policies could undermine confidence and trust on the part of developing countries that the Annex I countries were prepared to shoulder responsibilities for reducing greenhouse gas emissions. This loss of confidence could reduce investments and lower prospects for world economic growth.

B. Annex I countries act but developing countries remain cautious. Industrialised countries decide to implement national policies and measures as well as some internationally co-ordinated policies (including some projects implemented jointly and/or the use of tradeable permits within Annex I countries). However, they do not succeed in reaching a global agreement with the developing countries, either because Annex I countries cannot agree positions among themselves or because most developing countries do not agree to participate. Such an outcome would lead to high costs for CO_2 emissions abatement that would reduce prospects for economic growth.

C. All countries decide to co-operate in implementing CO_2 abatement policies. This would be the most successful outcome. It could combine the following features:

i) take advantage of lower-cost CO_2 abatement options in the developing countries, thus lowering the global cost of reducing CO_2 emissions (as compared with B above). If a system of activities implemented jointly combined with a system of tradeable permits were agreed and implemented, developing countries would be able to sell their emission reduction permits to industrialised countries, which, in turn, would benefit from lower abatement costs;

ii) develop trust and confidence between developing and developed countries that would promote globalisation, increased international trade and foreign direct investment, and thus potentially increase the rate of economic growth above that achievable in A above;

iii) provide the necessary market signals to spur the development and global diffusion of climate-friendly technologies.

In addition to the trust and confidence issue highlighted in C above, a number of other potentially positive factors for economic growth can be identified. These factors could offset the negative impact of higher costs on economic growth:

- possible removal of existing distortionary taxes, using revenues from climate change policies (where this is practicable);

- possible spin-offs from new CO_2 reduction technologies that generate extra growth;

- longer-term reductions in the unit costs of CO_2 emission reduction technologies;

- more available or cheaper investment financing for those with better abilities to locate and earn higher returns on investment than under a business-as-usual projection;

- enhanced energy security.

The limited analysis undertaken so far does not permit quantification of the costs and benefits and the economic growth impacts discussed in A, B and C above. Such analysis needs to be pursued.

SUMMARY

Climate-change policies agreed in Kyoto in 1997 could affect some sections of the world's capital stock of equipment, buildings and vehicles by 2010.

Technology development potentially offers the greatest scope in the longer term for reducing the unit costs of CO_2 emission abatement options.

A package of climate-change policies would be likely to reduce coal consumption and boost gas demand, as compared with a "business-as-usual" projection; oil demand may be little affected.

Nuclear power will be favoured in some countries, subject to public acceptability, safety, cost and environmental factors.

Renewable sources of energy, too, are likely to gain an advantage over fossil fuels under climate-change policies, but they will be held back by cost limitations and by constraints due to their visual and environmental impacts.

Fossil fuel reserve estimates are extremely uncertain. However, assuming that current consumption trends continue, estimates suggest that production of conventional oil could peak between 2000 and 2020. Natural gas supply is expected to be sufficient to meet growing demand, but production could peak before 2035.

Business-as-usual energy projections indicate that CO_2 concentrations will continue to rise and will eventually double the pre-industrial level unless substantial new and effective policies to abate CO_2 emissions are implemented.

All countries will need to co-operate to address the climate change threat. Annex I countries have agreed to take the lead by committing themselves to quantified emissions limitation or reduction. Developing countries already have reporting, inventory, and mitigation commitments. Joint Implementation, with credits accruing to countries on a mutually agreed basis, can provide a market-based monetary incentive to reduce emissions and to transfer efficient, climate-friendly technology to developing countries. Emissions would thus be reduced in the most cost-effective way, with developed countries bearing a substantial part of the incremental cost of the reductions.

MAIN SALES OUTLETS OF OECD PUBLICATIONS
PRINCIPAUX POINTS DE VENTE DES PUBLICATIONS DE L'OCDE

AUSTRALIA – AUSTRALIE
D.A. Information Services
648 Whitehorse Road, P.O.B 163
Mitcham, Victoria 3132 Tel. (03) 9210.7777
Fax: (03) 9210.7788

AUSTRIA – AUTRICHE
Gerold & Co.
Graben 31
Wien I Tel. (0222) 533.50.14
Fax: (0222) 512.47.31.29

BELGIUM – BELGIQUE
Jean De Lannoy
Avenue du Roi, Koningslaan 202
B-1060 Bruxelles Tel. (02) 538.51.69/538.08.41
Fax: (02) 538.08.41

CANADA
Renouf Publishing Company Ltd.
5369 Canotek Road
Unit 1
Ottawa, Ont. K1J 9J3 Tel. (613) 745.2665
Fax: (613) 745.7660
Stores:
71 1/2 Sparks Street
Ottawa, Ont. K1P 5R1 Tel. (613) 238.8985
Fax: (613) 238.6041
12 Adelaide Street West
Toronto, QN M5H 1L6 Tel. (416) 363.3171
Fax: (416) 363.5963
Les Éditions La Liberté Inc.
3020 Chemin Sainte-Foy
Sainte-Foy, PQ G1X 3V6 Tel. (418) 658.3763
Fax: (418) 658.3763
Federal Publications Inc.
165 University Avenue, Suite 701
Toronto, ON M5H 3B8 Tel. (416) 860.1611
Fax: (416) 860.1608
Les Publications Fédérales
1185 Université
Montréal, QC H3B 3A7 Tel. (514) 954.1633
Fax: (514) 954.1635

CHINA – CHINE
Book Dept., China Natinal Publiations
Import and Export Corporation (CNPIEC)
16 Gongti E. Road, Chaoyang District
Beijing 100020 Tel. (10) 6506-6688 Ext. 8402
(10) 6506-3101

CHINESE TAIPEI – TAIPEI CHINOIS
Good Faith Worldwide Int'l. Co. Ltd.
9th Floor, No. 118, Sec. 2
Chung Hsiao E. Road
Taipei Tel. (02) 391.7396/391.7397
Fax: (02) 394.9176

CZECH REPUBLIC – RÉPUBLIQUE TCHÈQUE
National Information Centre
NIS – prodejna
Konviktská 5
Praha 1 – 113 57 Tel. (02) 24.23.09.07
Fax: (02) 24.22.94.33
E-mail: nkposp@dec.niz.cz
Internet: http://www.nis.cz

DENMARK – DANEMARK
Munksgaard Book and Subscription Service
35, Nørre Søgade, P.O. Box 2148
DK-1016 København K Tel. (33) 12.85.70
Fax: (33) 12.93.87
J. H. Schultz Information A/S,
Herstedvang 12,
DK – 2620 Albertslung Tel. 43 63 23 00
Fax: 43 63 19 69
Internet: s-info@inet.uni-c.dk

EGYPT – ÉGYPTE
The Middle East Observer
41 Sherif Street
Cairo Tel. (2) 392.6919
Fax: (2) 360.6804

FINLAND – FINLANDE
Akateeminen Kirjakauppa
Keskuskatu 1, P.O. Box 128
00100 Helsinki

Subscription Services/Agence d'abonnements :
P.O. Box 23
00100 Helsinki Tel. (358) 9.121.4403
Fax: (358) 9.121.4450

***FRANCE**
OECD/OCDE
Mail Orders/Commandes par correspondance :
2, rue André-Pascal
75775 Paris Cedex 16 Tel. 33 (0)1.45.24.82.00
Fax: 33 (0)1.49.10.42.76
Telex: 640048 OCDE
Internet: Compte.PUBSINQ@oecd.org

Orders via Minitel, France only/
Commandes par Minitel, France exclusivement :
36 15 OCDE

OECD Bookshop/Librairie de l'OCDE :
33, rue Octave-Feuillet
75016 Paris Tel. 33 (0)1.45.24.81.81
33 (0)1.45.24.81.67
Dawson
B.P. 40
91121 Palaiseau Cedex Tel. 01.89.10.47.00
Fax: 01.64.54.83.26
Documentation Française
29, quai Voltaire
75007 Paris Tel. 01.40.15.70.00
Economica
49, rue Héricart
75015 Paris Tel. 01.45.78.12.92
Fax: 01.45.75.05.67
Gibert Jeune (Droit-Économie)
6, place Saint-Michel
75006 Paris Tel. 01.43.25.91.19
Librairie du Commerce International
10, avenue d'Iéna
75016 Paris Tel. 01.40.73.34.60
Librairie Dunod
Université Paris-Dauphine
Place du Maréchal-de-Lattre-de-Tassigny
75016 Paris Tel. 01.44.05.40.13
Librairie Lavoisier
11, rue Lavoisier
75008 Paris Tel. 01.42.65.39.95
Librairie des Sciences Politiques
30, rue Saint-Guillaume
75007 Paris Tel. 01.45.48.36.02
P.U.F.
49, boulevard Saint-Michel
75005 Paris Tel. 01.43.25.83.40
Librairie de l'Université
12a, rue Nazareth
13100 Aix-en-Provence Tel. 04.42.26.18.08
Documentation Française
165, rue Garibaldi
69003 Lyon Tel. 04.78.63.32.23
Librairie Decitre
29, place Bellecour
69002 Lyon Tel. 04.72.40.54.54
Librairie Sauramps
Le Triangle
34967 Montpellier Cedex 2 Tel. 04.67.58.85.15
Fax: 04.67.58.27.36
A la Sorbonne Actual
23, rue de l'Hôtel-des-Postes
06000 Nice Tel. 04.93.13.77.75
Fax: 04.93.80.75.69

GERMANY – ALLEMAGNE
OECD Bonn Centre
August-Bebel-Allee 6
D-53175 Bonn Tel. (0228) 959.120
Fax: (0228) 959.12.17

GREECE – GRÈCE
Librairie Kauffmann
Stadiou 28
10564 Athens Tel. (01) 32.55.321
Fax: (01) 32.30.320

HONG-KONG
Swindon Book Co. Ltd.
Astoria Bldg. 3F
34 Ashley Road, Tsimshatsui
Kowloon, Hong Kong Tel. 2376.2062
Fax: 2376.0685

HUNGARY – HONGRIE
Euro Info Service
Margitsziget, Európa Ház
1138 Budapest Tel. (1) 111.60.61
Fax: (1) 302.50.35
E-mail: euroinfo@mail.matav.hu
Internet: http://www.euroinfo.hu//index.html

ICELAND – ISLANDE
Mál og Menning
Laugavegi 18, Pósthólf 392
121 Reykjavik Tel. (1) 552.4240
Fax: (1) 562.3523

INDIA – INDE
Oxford Book and Stationery Co.
Scindia House
New Delhi 110001 Tel. (11) 331.5896/5308
Fax: (11) 332.2639
E-mail: oxford.publ@axcess.net.in
17 Park Street
Calcutta 700016 Tel. 240832

INDONESIA – INDONÉSIE
Pdii-Lipi
P.O. Box 4298
Jakarta 12042 Tel. (21) 573.34.67
Fax: (21) 573.34.67

IRELAND – IRLANDE
Government Supplies Agency
Publications Section
4/5 Harcourt Road
Dublin 2 Tel. 661.31.11
Fax: 475.27.60

ISRAEL – ISRAËL
Praedicta
5 Shatner Street
P.O. Box 34030
Jerusalem 91430 Tel. (2) 652.84.90/1/2
Fax: (2) 652.84.93
R.O.Y. International
P.O. Box 13056
Tel Aviv 61130 Tel. (3) 546 1423
Fax: (3) 546 1442
E-mail: royil@netvision.net.il
Palestinian Authority/Middle East:
INDEX Information Services
P.O.B. 19502
Jerusalem Tel. (2) 627.16.34
Fax: (2) 627.12.19

ITALY – ITALIE
Libreria Commissionaria Sansoni
Via Duca di Calabria, 1/1
50125 Firenze Tel. (055) 64.54.15
Fax: (055) 64.12.57
E-mail: licosa@ftbcc.it
Via Bartolini 29
20155 Milano Tel. (02) 36.50.83
Editrice e Libreria Herder
Piazza Montecitorio 120
00186 Roma Tel. 679.46.28
Fax: 678.47.51
Libreria Hoepli
Via Hoepli 5
20121 Milano Tel. (02) 86.54.46
Fax: (02) 805.28.86

Libreria Scientifica
Dott. Lucio de Biasio 'Aeiou'
Via Coronelli, 6
20146 Milano Tel. (02) 48.95.45.52
Fax: (02) 48.95.45.48

JAPAN – JAPON
OECD Tokyo Centre
Landic Akasaka Building
2-3-4 Akasaka, Minato-ku
Tokyo 107 Tel. (81.3) 3586.2016
Fax: (81.3) 3584.7929

KOREA – CORÉE
Kyobo Book Centre Co. Ltd.
P.O. Box 1658, Kwang Hwa Moon
Seoul Tel. 730.78.91
Fax: 735.00.30

MALAYSIA – MALAISIE
University of Malaya Bookshop
University of Malaya
P.O. Box 1127, Jalan Pantai Baru
59700 Kuala Lumpur
Malaysia Tel. 756.5000/756.5425
Fax: 756.3246

MEXICO – MEXIQUE
OECD Mexico Centre
Edificio INFOTEC
Av. San Fernando no. 37
Col. Toriello Guerra
Tlalpan C.P. 14050
Mexico D.F. Tel. (525) 528.10.38
Fax: (525) 606.13.07
E-mail: ocde@rtn.net.mx

NETHERLANDS – PAYS-BAS
SDU Uitgeverij Plantijnstraat
Externe Fondsen
Postbus 20014
2500 EA's-Gravenhage Tel. (070) 37.89.880
Voor bestellingen: Fax: (070) 34.75.778
Subscription Agency/ Agence d'abonnements :
SWETS & ZEITLINGER BV
Heereweg 347B
P.O. Box 830
2160 SZ Lisse Tel. 252.435.111
Fax: 252.415.888

NEW ZEALAND – NOUVELLE-ZÉLANDE
GPLegislation Services
P.O. Box 12418
Thorndon, Wellington Tel. (04) 496.5655
Fax: (04) 496.5698

NORWAY – NORVÈGE
NIC INFO A/S
Ostensjoveien 18
P.O. Box 6512 Etterstad
0606 Oslo Tel. (22) 97.45.00
Fax: (22) 97.45.45

PAKISTAN
Mirza Book Agency
65 Shahrah Quaid-E-Azam
Lahore 54000 Tel. (42) 735.36.01
Fax: (42) 576.37.14

PHILIPPINE – PHILIPPINES
International Booksource Center Inc.
Rm 179/920 Cityland 10 Condo Tower 2
HV dela Costa Ext cor Valero St.
Makati Metro Manila Tel. (632) 817 9676
Fax: (632) 817 1741

POLAND – POLOGNE
Ars Polona
00-950 Warszawa
Krakowskie Prezdmiescie 7 Tel. (22) 264760
Fax: (22) 265334

PORTUGAL
Livraria Portugal
Rua do Carmo 70-74
Apart. 2681
1200 Lisboa Tel. (01) 347.49.82/5
Fax: (01) 347.02.64

SINGAPORE – SINGAPOUR
Ashgate Publishing
Asia Pacific Pte. Ltd
Golden Wheel Building, 04-03
41, Kallang Pudding Road
Singapore 349316 Tel. 741.5166
Fax: 742.9356

SPAIN – ESPAGNE
Mundi-Prensa Libros S.A.
Castelló 37, Apartado 1223
Madrid 28001 Tel. (91) 431.33.99
Fax: (91) 575.39.98
E-mail: mundiprensa@tsai.es
Internet: http://www.mundiprensa.es

Mundi-Prensa Barcelona
Consell de Cent No. 391
08009 – Barcelona Tel. (93) 488.34.92
Fax: (93) 487.76.59

Libreria de la Generalitat
Palau Moja
Rambla dels Estudis, 118
08002 – Barcelona
(Suscripciones) Tel. (93) 318.80.12
(Publicaciones) Tel. (93) 302.67.23
Fax: (93) 412.18.54

SRI LANKA
Centre for Policy Research
c/o Colombo Agencies Ltd.
No. 300-304, Galle Road
Colombo 3 Tel. (1) 574240, 573551-2
Fax: (1) 575394, 510711

SWEDEN – SUÈDE
CE Fritzes AB
S-106 47 Stockholm Tel. (08) 690.90.90
Fax: (08) 20.50.21

For electronic publications only/
Publications électroniques seulement
STATISTICS SWEDEN
Informationsservice
S-115 81 Stockholm Tel. 8 783 5066
Fax: 8 783 4045

Subscription Agency/Agence d'abonnements :
Wennergren-Williams Info AB
P.O. Box 1305
171 25 Solna Tel. (08) 705.97.50
Fax: (08) 27.00.71

Liber distribution
International organizations
Fagerstagatan 21
S-163 52 Spanga

SWITZERLAND – SUISSE
Maditec S.A. (Books and Periodicals/Livres
et périodiques)
Chemin des Palettes 4
Case postale 266
1020 Renens VD 1 Tel. (021) 635.08.65
Fax: (021) 635.07.80

Librairie Payot S.A.
4, place Pépinet
CP 3212
1002 Lausanne Tel. (021) 320.25.11
Fax: (021) 320.25.14

Librairie Unilivres
6, rue de Candolle
1205 Genève Tel. (022) 320.26.23
Fax: (022) 329.73.18

Subscription Agency/Agence d'abonnements :
Dynapresse Marketing S.A.
38, avenue Vibert
1227 Carouge Tel. (022) 308.08.70
Fax: (022) 308.07.99

See also – Voir aussi :
OECD Bonn Centre
August-Bebel-Allee 6
D-53175 Bonn (Germany) Tel. (0228) 959.120
Fax: (0228) 959.12.17

THAILAND – THAÏLANDE
Suksit Siam Co. Ltd.
113, 115 Fuang Nakhon Rd.
Opp. Wat Rajbopith
Bangkok 10200 Tel. (662) 225.9531/2
Fax: (662) 222.5188

**TRINIDAD & TOBAGO, CARIBBEAN
TRINITÉ-ET-TOBAGO, CARAÏBES**
Systematics Studies Limited
9 Watts Street
Curepe
Trinadad & Tobago, W.I. Tel. (1809) 645.3475
Fax: (1809) 662.5654
E-mail: tobe@trinidad.net

TUNISIA – TUNISIE
Grande Librairie Spécialisée
Fendri Ali
Avenue Haffouz Imm El-Intilaka
Bloc B 1 Sfax 3000 Tel. (216-4) 296 855
Fax: (216-4) 298.270

TURKEY – TURQUIE
Kültür Yayinlari Is-Türk Ltd.
Atatürk Bulvari No. 191/Kat 13
06684 Kavaklidere/Ankara
Tel. (312) 428.11.40 Ext. 2458
Fax : (312) 417.24.90

Dolmabahce Cad. No. 29
Besiktas/Istanbul Tel. (212) 260 7188

UNITED KINGDOM – ROYAUME-UNI
The Stationery Office Ltd.
Postal orders only:
P.O. Box 276, London SW8 5DT
Gen. enquiries Tel. (171) 873 0011
Fax: (171) 873 8463

The Stationery Office Ltd.
Postal orders only:
49 High Holborn, London WC1V 6HB
Branches at: Belfast, Birmingham, Bristol,
Edinburgh, Manchester

UNITED STATES – ÉTATS-UNIS
OECD Washington Center
2001 L Street N.W., Suite 650
Washington, D.C. 20036-4922 Tel. (202) 785.6323
Fax: (202) 785.0350
Internet: washcont@oecd.org

Subscriptions to OECD periodicals may also be
placed through main subscription agencies.

Les abonnements aux publications périodiques de
l'OCDE peuvent être souscrits auprès des
principales agences d'abonnement.

Orders and inquiries from countries where Distributors have not yet been appointed should be sent to:
OECD Publications, 2, rue André-Pascal, 75775
Paris Cedex 16, France.

Les commandes provenant de pays où l'OCDE n'a
pas encore désigné de distributeur peuvent être
adressées aux Éditions de l'OCDE, 2, rue André-Pascal, 75775 Paris Cedex 16, France.

12-1996

OECD PUBLICATIONS, 2 rue André-Pascal, 75775 PARIS CEDEX 16
PRINTED IN FRANCE
(61 97 31 1 P) ISBN 92-64-15668-2 - No. 49779 1997